Graph Data Science with Neo4j

Learn how to use Neo4j 5 with Graph Data Science library 2.0 and its Python driver for your project

Estelle Scifo

BIRMINGHAM—MUMBAI

Graph Data Science with Neo4j

Publishing Product Manager: Ali Abidi
Senior Editor: Nathanya Dias
Technical Editor: Rahul Limbachiya
Copy Editor: Safis Editing
Project Coordinator: Farheen Fathima
Proofreader: Safis Editing
Indexer: Hemangini Bari
Production Designer: Shankar Kalbhor
Marketing Coordinator: Vinishka Kalra

First published: January 2023

Production reference: 1310123

Published by Packt Publishing Ltd.
Livery Place
35 Livery Street
Birmingham
B3 2PB, UK.

ISBN 978-1-80461-274-3

www.packtpub.com

Contributors

About the author

Estelle Scifo is a Neo4j Certified Professional and Neo4j Graph Data Science certified user. She is currently a machine learning engineer at GraphAware where she builds Neo4j-related solutions to make customers happy with graphs.

Before that, she worked in several fields, starting out with research in particle physics, during which she worked at CERN on uncovering Higgs boson properties. She received her PhD in 2014 from the Laboratoire de l'Accélérateur Linéaire (Orsay, France). Continuing her career in industry, she worked in real estate, mobility, and logistics for almost 10 years. In the Neo4j community, she is known as the creator of neomap, a map visualization application for data stored in Neo4j. She also regularly gives talks at conferences such as NODES and PyCon. Her domain expertise and deep insight into the perspective of a beginner's needs make her an excellent teacher.

There is only one name on the cover, but a book is not the work of one person. I would like to thank everyone involved in making this book a reality. Beyond everyone at Packt, the reviewers did an incredible job of suggesting some very relevant improvements. Thank you, all!

I hope this book will inspire you as much as other books of this genre have inspired me.

About the reviewers

Dr. David Gurzick is the founding chair of the George B. Delaplaine Jr. School of Business and an associate professor of management science at Hood College. He has a BS in computer science from Frostburg State University, an M.S. in computer science from Hood College, a PhD in information systems from the University of Maryland, Baltimore County, and is a graduate of Harvard's business analytics program. As a child of the internet, he grew up on AOL and programmed his way through dot.com. He now helps merge technology and business strategy to enable innovation and accelerate commercial success as the lead data scientist at Genitive.ai and as a director of the **Frederick Innovative Technology Center, Inc (FITCI)**.

Sean William Grant is a product and analytics professional with over 20 years of experience in technology and data analysis. His experience ranges from geospatial intelligence with the United States Marine Corps, product management within the aviation and autonomy space, to implementing advanced analytics and data science within organizations. He is a graph data science and network analytics enthusiast who frequently gives presentations and workshops on connected data. He has also been a technical advisor to several early-stage start-ups. Sean is passionate about data and technology, and how it can elevate our understanding of ourselves.

Jose Ernesto Echeverria has worked with all kinds of databases, from relational databases in the 1990s to non-SQL databases in the 2010s. He considers graph databases to be the best fit for solving real-world problems, given their strong capability for modeling and adaptability to change. As a polyglot programmer, he has used languages such as Java, Ruby, and R and tools such as Jupyter with Neo4j in order to solve data management problems for multinational corporations. A long-time advocate of data science, he expects this long-awaited book to cover the proper techniques and approach the intersections of this discipline, as well as help readers to discover the possibilities of graph databases. When not working, he enjoys spending time with friends and family.

Table of Contents

7

Automatically Extracting Features with Graph Embeddings for Machine Learning 177

8

Building a GDS Pipeline for Node Classification Model Training 201

9

Predicting Future Edges 223

10

Writing Your Custom Graph Algorithms with the Pregel API in Java 237

Index 259

Other Books You May Enjoy 268

Preface

Data science today is a core component of many companies and organizations taking advantage of its predictive power to improve their products or better understand their customers. It is an ever-evolving field, still undergoing intense research. One of the most trending research areas is **graph data science** (**GDS**), or how representing data as a connected network can improve models.

Among the different tools on the market to work with graphs, Neo4j, a graph database, is popular among developers for its ability to build simple and evolving data models and query data easily with Cypher. For a few years now, it has also stood out as a leader in graph analytics, especially since the release of the first version of its GDS library, allowing you to run graph algorithms from data stored in Neo4j, even at a large scale.

This book is designed to guide you through the field of GDS, always using Neo4j and its GDS library as the main tool. By the end of this book, you will be able to run your own *GDS model* on a graph dataset you created. By the end of the book, you will even be able to pass the Neo4j Data Science certification to prove your new skills to the world.

Who this book is for

This book is for people who are curious about graphs and how this data structure can be useful in data science. It can serve both data scientists who are learning about graphs and Neo4j developers who want to get into data science.

The book assumes minimal data science knowledge (classification, training sets, confusion matrices) and some experience with Python and its related data science toolkit (pandas, matplotlib, and scikit-learn).

What this book covers

Chapter 1, Introducing and Installing Neo4j, introduces the basic principles of graph databases and gives instructions on how to set up Neo4j locally, create your first graph, and write your first Cypher queries.

Chapter 2, Using Existing Data to Build a Knowledge Graph, guides you through loading data into Neo4j from different formats (CSV, JSON, and an HTTP API). This is where you will build the dataset that will be used throughout this book.

Chapter 3, Characterizing a Graph Dataset, introduces some key metrics to differentiate one graph dataset from another.

Chapter 4, Using Graph Algorithms to Characterize a Graph Dataset, goes deeper into understanding a graph dataset by using graph algorithms. This is the chapter where you will start to use the Neo4j GDS plugin.

Chapter 5, Visualizing Graph Data, delves into graph data visualization by drawing nodes and edges, starting from static representations and moving on to dynamic ones.

Chapter 6, Building a Machine Learning Model with Graph Features, talks about machine learning model training using scikit-learn. This is where we will first use the GDS Python client.

Chapter 7, Automating Feature Extraction with Graph Embeddings for Machine Learning, introduces the concept of node embedding, with practical examples using the Neo4j GDS library.

Chapter 8, Building a GDS Pipeline for Node Classification Model Training, introduces the topic of node classification within GDS without involving a third-party tool.

Chapter 9, Predicting Future Edges, gives a short introduction to the topic of link prediction, a graph-specific machine learning task.

Chapter 10, Writing Your Custom Graph Algorithms with the Pregel API in Java, covers the exciting topic of building an extension for the GDS plugin.

To get the most out of this book

You will need access to a Neo4j instance. Options and installation instructions are given in *Chapter 1, Introducing and Installing Neo4j*. We will also intensively use Python and the following packages: pandas, scikit-learn, network, and graphdatascience. The code was tested with Python 3.10 but should work with newer versions, assuming no breaking change is made in its dependencies. Python code is provided as a Jupyter notebook, so you'll need Jupyter Server installed and running to go through it.

For the very last chapter, a Java JDK will also be required. The code was tested with OpenJDK 11.

Software/hardware covered in the book	Operating system requirements
Neo4j 5.x	Windows, macOS, or Linux
Python 3.10	Windows, macOS or Linux
Jupyter	Windows, macOS or Linux
OpenJDK 11	Windows, macOS or Linux

You will also need to install Neo4j plugins: APOC and GDS. Installation instructions for Neo4j Desktop are given in the relevant chapters. However, if you are not using a local Neo4j instance, please refer to the following pages for installation instructions, especially regarding version compatibilities:

- *APOC*: `https://neo4j.com/docs/apoc/current/installation/`
- *GDS*: `https://neo4j.com/docs/graph-data-science/current/installation/`

If you are using the digital version of this book, we advise you to type the code yourself or access the code from the book's GitHub repository (a link is available in the next section). Doing so will help you avoid any potential errors related to the copying and pasting of code.

Download the example code files

You can download the example code files for this book from GitHub at `https://github.com/PacktPublishing/Graph-Data-Science-with-Neo4j`. If there's an update to the code, it will be updated in the GitHub repository.

We also have other code bundles from our rich catalog of books and videos available at `https://github.com/PacktPublishing/`. Check them out!

Conventions used

There are a number of text conventions used throughout this book.

`Code in text`: Indicates code words in text, database table names, folder names, filenames, file extensions, pathnames, dummy URLs, user input, and Twitter handles. Here is an example: "Mount the downloaded `WebStorm-10*.dmg` disk image file as another disk in your system."

A block of code is set as follows:

```
CREATE (:Movie {
    id: line.show_id,
    title: line.title,
    releaseYear: line.release_year
  }
```

When we wish to draw your attention to a particular part of a code block, the relevant lines or items are set in bold:

```
LOAD CSV WITH HEADERS
FROM 'file:///netflix/netflix_titles.csv' AS line
WITH split(line.director, ",") as directors_list
UNWIND directors_list AS director_name
CREATE (:Person {name: trim(director_name)})
```

Any command-line input or `output` is written as follows:

```
$ mkdir css
$ cd css
```

Bold: Indicates a new term, an important word, or words that you see onscreen. For instance, words in menus or dialog boxes appear in **bold**. Here is an example: "Select **System info** from the **Administration** panel."

> **Tips or important notes**
> Appear like this.

Get in touch

Feedback from our readers is always welcome.

General feedback: If you have questions about any aspect of this book, email us at `customercare@packtpub.com` and mention the book title in the subject of your message.

Errata: Although we have taken every care to ensure the accuracy of our content, mistakes do happen. If you have found a mistake in this book, we would be grateful if you would report this to us. Please visit `www.packtpub.com/support/errata` and fill in the form.

Piracy: If you come across any illegal copies of our works in any form on the internet, we would be grateful if you would provide us with the location address or website name. Please contact us at `copyright@packt.com` with a link to the material.

If you are interested in becoming an author: If there is a topic that you have expertise in and you are interested in either writing or contributing to a book, please visit `authors.packtpub.com`.

Share Your Thoughts

Once you've read *Graph Data Science with Neo4j*, we'd love to hear your thoughts! Scan the QR code below to go straight to the Amazon review page for this book and share your feedback.

`https://packt.link/r/1-804-61274-X`

Your review is important to us and the tech community and will help us make sure we're delivering excellent quality content.

Download a free PDF copy of this book

Thanks for purchasing this book!

Do you like to read on the go but are unable to carry your print books everywhere? Is your eBook purchase not compatible with the device of your choice?

Don't worry, now with every Packt book you get a DRM-free PDF version of that book at no cost.

Read anywhere, any place, on any device. Search, copy, and paste code from your favorite technical books directly into your application.

The perks don't stop there, you can get exclusive access to discounts, newsletters, and great free content in your inbox daily

Follow these simple steps to get the benefits:

1. Scan the QR code or visit the link below

`https://packt.link/free-ebook/9781804612743`

2. Submit your proof of purchase
3. That's it! We'll send your free PDF and other benefits to your email directly

Part 1 – Creating Graph Data in Neo4j

In this first part, you will learn about Neo4j and set up your first graph database. You will also build a graph dataset in Neo4j using Cypher, the APOC library, and public knowledge graphs.

This part includes the following chapters:

1
Introducing and Installing Neo4j

Graph databases in general, and Neo4j in particular, have gained increasing interest in the past few years. They provide a natural way of modeling entities and relationships and take into account observation context, which is often crucial to extract the most out of your data. Among the different graph database vendors, Neo4j has become one of the most popular for both data storage and analytics. A lot of tools have been developed by the company itself or the community to make the whole ecosystem consistent and easy to use: from storage to querying, to visualization to graph data science. As you will see through this book, there is a well-integrated application or plugin for each of these topics.

In this chapter, you will get to know what Neo4j is, positioning it in the broad context of databases. We will also introduce the aforementioned plugins that are used for graph data science.

Finally, you will set up your first Neo4j instance locally if you haven't done so already and run your first Cypher queries to populate the database with some data and retrieve it.

In this chapter, we're going to cover the following main topics:

- What is a graph database?
- Finding or creating a graph database
- Neo4j in the graph databases landscape
- Setting up Neo4j
- Inserting data into Neo4j with Cypher, the Neo4j query language
- Extracting data from Neo4j with Cypher pattern matching

Technical requirements

To follow this chapter well, you will need access to the following resources:

- You'll need a computer that can run Neo4j locally; Windows, macOS, and Linux are all supported. Please refer to the Neo4j website for more details about system requirements: `https://neo4j.com/docs/operations-manual/current/installation/requirements/`.
- Any code listed in the book will be available in the associated GitHub repository – that is, `https://github.com/PacktPublishing/Graph-Data-Science-with-Neo4j` – in the corresponding chapter folder.

What is a graph database?

Before we get our hands dirty and start playing with Neo4j, it is important to understand what Neo4j is and how different it is from the data storage engine you are used to. In this section, we are going to discuss (quickly) the different types of databases you can find today, and why graph databases are so interesting and popular both for developers and data professionals.

Databases

Databases make up an important part of computer science. Discussing the evolution and state-of-the-art areas of the different implementations in detail would require several books like this one – fortunately, this is not a requirement to use such systems effectively. However, it is important to be aware of the existing tools related to data storage and how they differ from each other, to be able to choose the right tool for the right task. The fact that, after reading this book, you'll be able to use graph databases and Neo4j in your data science project doesn't mean you will have to use it every single time you start a new project, whatever the context is. Sometimes, it won't be suitable; this introduction will explain why.

A database, in the context of computing, is a system that allows you to store and query data on a computer, phone, or, more generally, any electronic device.

As developers or data scientists of the 2020s, we have mainly faced two kinds of databases:

- **Relational databases** (**SQL**) such as MySQL or PostgreSQL. These store data as records in tables whose columns are attributes or fields and whose rows represent each entity. They have a predefined schema, defining how data is organized and the type of each field. Relationships between entities in this representation are modeled by foreign keys (requiring unique identifiers). When the relationship is more complex, such as when attributes are required or when we can have many relationships between the same objects, an intermediate *junction* (join) table is required.

- **NoSQL databases** contain many different types of databases:
  - **Key-value stores** such as Redis or Riak. A **key-value** (**KV**) store, as the name suggests, is a simple lookup database where the key is usually a string, and the value can be a more complex object that can't be used to filter the query – it can only be retrieved. They are known to be very efficient for caching in a web context, where the key is the page URL and the value is the HTML content of the page, which is dynamically generated. KV stores can also be used to model graphs when building a native graph engine is not an option. You can see KV stores in action in the following projects:
    - **IndraDB**: This is a graph database written in Rust that relies on different types of KV stores: `https://github.com/indradb/indradb`
  - **Document-oriented databases** such as MongoDB or CouchDB. These are useful for storing schema-less documents (usually JSON (or a derivative) objects). They are much more flexible compared to relational databases, since each document may have different fields. However, relationships are harder to model, and such databases rely a lot on nested JSON and information duplication instead of joining multiple tables.

The preceding list is non-exhaustive; other types of data stores have been created over time and abandoned or were born in the past years, so we'll need to wait to see how useful they can be. We can mention, for instance, vector databases, such as Weaviate, which are used to store data with their vector representations to ease searching in the vector space, with many applications in machine learning once a vector representation (embedding) of an observation has been computed.

Graph databases can also be classified as NoSQL databases. They bring another approach to the data storage landscape, especially in the data model phase.

Graph database

In the previous section, we talked about databases. Before discussing graph databases, let's introduce the concept of graphs.

A graph is a mathematical object defined by the following:

- A set of vertices or nodes (the dots)
- A set of edges (the connections between these dots)

The following figure shows several examples of graphs, big and small:

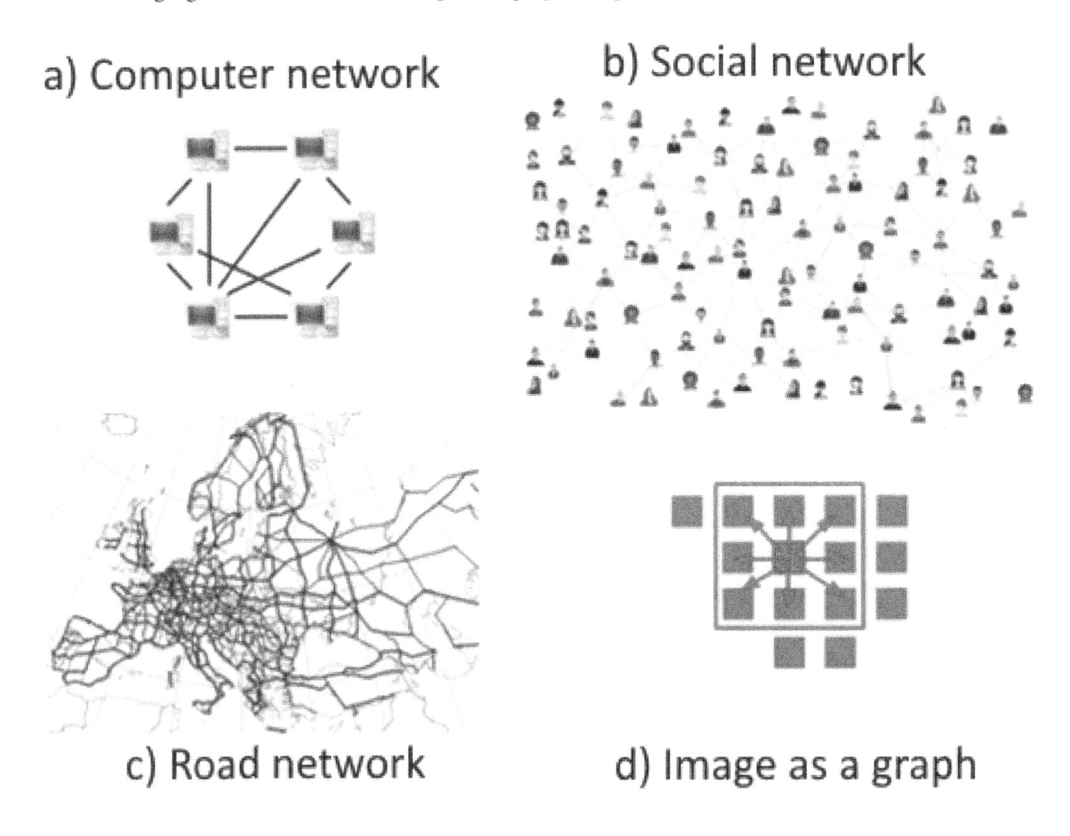

Figure 1.1 – Representations of some graphs

As you can see, there's a **Road network** (in Europe), a **Computer network**, and a **Social network**. But in practice, far more objects can be seen as graphs:

- **Time series**: Each observation is connected to the next one
- **Images**: Each pixel is linked to its eight neighbors (see the bottom-right picture in *Figure 1.1*)
- **Texts**: Here, each word is connected to its surrounding words or a more complex mapping, depending on its meaning (see the following figure):

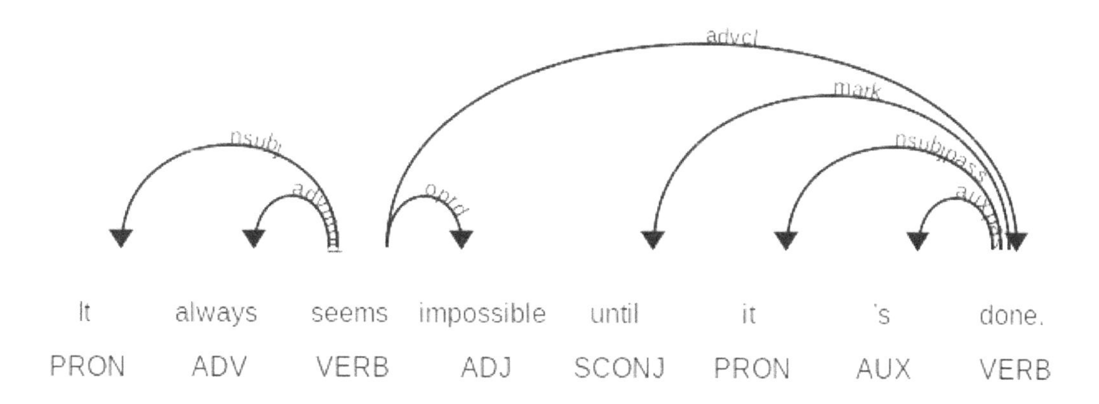

Figure 1.2 – Figure generated with the spacy Python library, which was able to identify the relationships between words in a sentence using NLP techniques

A graph can be seen as a generalization of these static representations, where links can be created with fewer constraints.

Another advantage of graphs is that they can be easily traversed, going from one node to another by following edges. They have been used for representing networks for a long time – road networks or communication infrastructure, for instance. The concept of a path, especially the shortest path in a graph, is a long-studied field. But the analysis of graphs doesn't stop here – much more information can be extracted from carefully analyzing a network, such as its structure (are there groups of nodes disconnected from the rest of the graph? Are groups of nodes more tied to each other than to other groups?) and node ranking (node importance). We will discuss these algorithms in more detail in *Chapter 4, Using Graph Algorithms to Characterize a Graph Dataset.*

So, we know what a database is and what a graph is. Now comes the natural question: what is a graph database? The answer is quite simple: in a graph database, data is saved into nodes, which can be connected through edges to model relationships between them.

At this stage, you may be wondering: ok, but where can I find graph data? While we are used to CSV or JSON datasets, graph formats are not yet common and it might be misleading to some of you. If you do not have graph data, why would you need a graph database? There are two possible answers to this question, both of which we are going to discuss.

Finding or creating a graph database

Data scientists know how to find or generate datasets that fit their needs. Randomly generating a variable distribution while following some probabilistic law is one of the first things you'll learn in a statistics course. Similarly, graph datasets can be randomly generated, following some rules. However, this book is not a graph theory book, so we are not going to dig into these details here. Just be aware that this can be done. Please refer to the references in the *Further reading* section to learn more.

Regarding existing datasets, some of them are very popular and data scientists know about them because they have used them while learning data science and/or because they are the topic of well-known Kaggle competitions. Think, for instance, about the *Titanic* or *house price* datasets. Other datasets are also used for model benchmarking, such as the *MNIST* or *ImageNet* datasets in computer vision tasks.

The same holds for graph data science, where some datasets are very common for teaching or benchmarking purposes. If you investigate graph theory, you will read about the **Zachary's karate club** (**ZKC**) dataset, which is probably one of the most famous graph datasets out there (side note: there is even a ZKC trophy, which is awarded to the first person in a graph conference that mentions this dataset). The ZKC dataset is very simple (30 nodes, as we'll see in *Chapter 3*, *Characterizing a Graph Dataset*, and *Chapter 4*, *Using Graph Algorithms to Characterize a Graph Dataset*, on how to characterize a graph dataset), but bigger and more complex datasets are also available.

There are websites referencing graph datasets, which can be used for benchmarking in a research context or educational purpose, such as this book. Two of the most popular ones are the following:

- The **Stanford Network Analysis Project** (**SNAP**) (`https://snap.stanford.edu/data/index.html`) lists different types of networks in different categories (social networks, citation networks, and so on)
- The **Network Repository Project**, via its website at `https://networkrepository.com/index.php`, provides hundreds of graph datasets from real-world examples, classified into categories (for example, biology, economics, recommendations, road, and so on)

If you browse these websites and start downloading some of the files, you'll notice the data comes in unfamiliar formats. We're going to list some of them next.

A note about the graph dataset's format

The datasets we are used to are mainly exchanged as CSV or JSON files. To represent a graph, with nodes on one side and edges on the other, several specific formats have been imagined.

The main data formats that are used to save graph data as text files are the following:

- **Edge list**: This is a text file where each row contains an edge definition. For instance, a graph with three nodes (`A`, `B`, `C`) and two edges (`A-B` and `C-A`) is defined by the following `edgelist` file:

```
A B
C A
```

- **Matrix Market** (with the `.mtx` extension): This format is an extension of the previous one. It is quite frequent on the network repository website.

- **Adjacency matrix**: The adjacency matrix is an `NxN` matrix (where `N` is the number of nodes in the graph) where the `ij` element is `1` if nodes `i` and `j` are connected through an edge and `0` otherwise. The adjacency matrix of the simple graph with three nodes and two edges is a `3x3` matrix, as shown in the following code block. I have explicitly displayed the row and column names only for convenience, to help you identify what `i` and `j` are:

```
  A B C
A 0 1 0
B 0 0 0
C 1 0 0
```

> **Note**
>
> The adjacency matrix is one way to *vectorize* a graph. We'll come back to this topic in *Chapter 7, Automatically Extracting Features with Graph Embeddings for Machine Learning.*

- **GraphML**: Derived from XML, the GraphML format is much more verbose but lets us define more complex graphs, especially those where nodes and/or edges carry properties. The following example uses the preceding graph but adds a `name` property to nodes and a `length` property to edges:

```
<?xml version='1.0' encoding='utf-8'?>
<graphml xmlns="http://graphml.graphdrawing.org/xmlns"
   xmlns:xsi="http://www.w3.org/2001/XMLSchema-instance"
   xsi:schemaLocation="http://graphml.graphdrawing.org/
xmlns http://graphml.graphdrawing.org/xmlns/1.0/graphml.
xsd"
>
    <!-- DEFINING PROPERTY NAME WITH TYPE AND ID -->
    <key attr.name="name" attr.type="string" for="node"
id="d1"/>
    <key attr.name="length" attr.type="double" for="edge"
id="d2"/>
    <graph edgedefault="directed">
       <!-- DEFINING NODES -->
       <node id="A">
             <!-- SETTING NODE PROPERTY -->
            <data key="d1">"Point A"</data>
        </node>
        <node id="B">
```

```
            <data key="d1">"Point B"</data>
        </node>
        <node id="C">
            <data key="d1">"Point C"</data>
        </node>
        <!-- DEFINING EDGES
        with source and target nodes and properties
    -->
        <edge id="AB" source="A" target="B">
            <data key="d2">123.45</data>
        </edge>
        <edge id="CA" source="C" target="A">
            <data key="d2">56.78</data>
        </edge>
    </graph>
</graphml>
```

If you find a dataset already formatted as a graph, it is likely to be using one of the preceding formats. However, most of the time, you will want to use your own data, which is not yet in graph format – it might be stored in the previously described databases or CSV or JSON files. If that is the case, then the next section is for you! There, you will learn how to transform your data into a graph.

Modeling your data as a graph

The second answer to the main question in this section is: your data is probably a graph, without you being aware of it yet. We will elaborate on this topic in the next chapter (*Chapter 2, Using Existing Data to Build a Knowledge Graph*), but let me give you a quick overview.

Let's take the example of an e-commerce website, which has customers (users) and products. As in every e-commerce website, users can place orders to buy some products. In the relational world, the data schema that's traditionally used to represent such a scenario is represented on the left-hand side of the following screenshot:

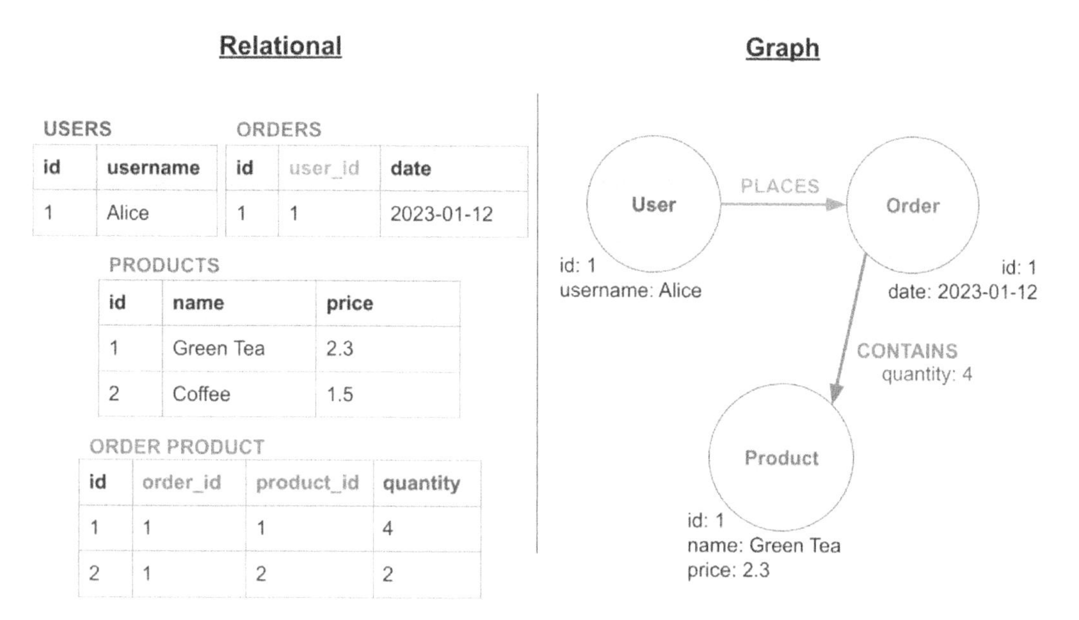

Figure 1.3 – Modeling e-commerce data as a graph

The relational data model works as follows:

- A table is created to store users, with a **unique identifier** (`id`) and a username (apart from security and personal information required for such a website, you can easily imagine how to add columns to this table).
- Another table contains the data about the available products.
- Each time a customer places an order, a new row is added to an order table, referencing the user by its ID (a foreign key with a one-to-many relationship, where a user can place many orders).
- To remember which products were part of which orders, a many-to-many relationship is created (an order contains many products and a product is part of many orders). We usually create a *relationship* table, linking orders to products (the order product table, in our example).

> **Note**
>
> Please refer to the colored version of the preceding figure, which can be found in the graphics bundle link provided in the *Preface*, for a better understanding of the correspondence between the two sides of the figure.

In a graph database, all the `_id` columns are replaced by actual *relationships*, which are real entities with graph databases, not just conceptual ones like in the relational model. You can also get rid of the `order product` table since information specific to a product in a given order such as the ordered quantity can be stored directly in the relationship between the order and the product node. The data model is much more natural and easier to document and present to other people on your team.

Now that we have a better understanding of what a graph database is, let's explore the different implementations out there. Like the other types of databases, there is no single implementation for graph databases, and several projects provide graph database functionalities.

In the next section, we are going to discuss some of the differences between them, and where Neo4j is positioned in this technology landscape.

Neo4j in the graph databases landscape

Even when restricting the scope to graph databases, there are still different ways to envision such data stores:

- **Resource description framework** (**RDF**): Each record is a triplet of the `Subject Predicate Object` type. This is a complex vocabulary that expresses a relationship of a certain type (the predicate) between a subject and an object; for instance:

  ```
  Alice(Subject) KNOWS(Predicate) Bob(Object)
  ```

  Very famous knowledge bases such as DBedia and Wikidata use the RDF format. We will talk about this a bit more in the next chapter (*Chapter 2, Using Existing Data to Build a Knowledge Graph*).

- **Labeled-property graph** (**LPG**): A labeled-property graph contains nodes and relationships. Both of these entities can be labeled (for instance, `Alice` and `Bob` are nodes with the `Person` label, and the relationship between them has the `KNOWS` label) and have properties (people have names; an acquaintance relationship can contain the date when both people first met as a property).

Neo4j is a labeled-property graph. And even there, like MySQL, PostgreSQL, and Microsoft SQL Server are all relational databases, you will find different vendors proposing LPG graph databases. They differ in many aspects:

- **Whether they use a native graph engine or not**: As we discussed earlier, it is possible to use a KV store or even a SQL database to store graph data. In this case, we're talking about non-native storage engines since the storage does not reflect the graphical nature of the data.

- **The query language**: Unlike SQL, the query language to deal with graph data has not yet been standardized, even if there is an ongoing effort being led by the GQL group (see, for instance, `https://gql.today/`). Neo4j uses Cypher, a declarative query language developed by the company in 2011 and then open-sourced in the `openCypher` project, allowing other databases to use the same language (see, for instance, RedisGraph or Amazon Neptune). Other vendors have created their own languages (`AQL` for ArangoDB or `CQL` for TigerGraph, for instance). To me, this is a key point to take into account since the learning curve can be very different from one language to another. Cypher has the advantage of being very intuitive and a few minutes are enough to write your own queries without much effort.

- Their (integrated or not) support for graph analytics and data science.

A note about performances

Almost every vendor claims to be the best one, at least in some aspects. This book won't create another debate about that. The best option, if performances are crucial for your application, is to test the candidates with a scenario close to your final goal in terms of data volume and the type of queries/analysis.

Neo4j ecosystem

The Neo4j database is already very helpful by itself, but the amount of extensions, libraries, and applications related to it makes it the most complete solution. In addition, it has a very active community of members always keen to help each other, which is one of the reasons to choose it.

The core Neo4j database capabilities can be extended thanks to some plugins. **Awesome Procedures on Cypher** (**APOC**), a common Neo4j extension, contains some procedures that can extend the database and Cypher capabilities. We will use it later in this book to load JSON data.

The main plugin we will explore in this book is the **Graph Data Science Library**. Its predecessor, the Graph Algorithm Library, was first released in 2018 by the Neo4j lab team. It was quickly replaced by the Graph Data Science Library, a fully production-ready plugin, with improved performance. Algorithms are improved and added regularly. Version 2.0, released in 2021, takes graph data science even further, allowing us to train models and build analysis pipelines directly from the library. It also comes with a handy Python client, which is very convenient for including graph algorithms into your usual machine learning processes, whether you use `scikit-learn` or other machine learning libraries such as TensorFlow or PyTorch.

Besides the plugins, there are also lots of applications out there to help us deal with Neo4j and explore the data it contains. The first application we will use is Neo4j Desktop, which lets us manage several Neo4j databases. Continue reading to learn how to use it. Neo4j Desktop also lets you manage your installed plugins and applications.

Applications installed into Neo4j Desktop are granted access to your active database. While reading this book, you will use the following:

- **Neo4j Browser**: A simple but powerful application that lets you write Cypher queries and visualize the result as a graph, table, or JSON:

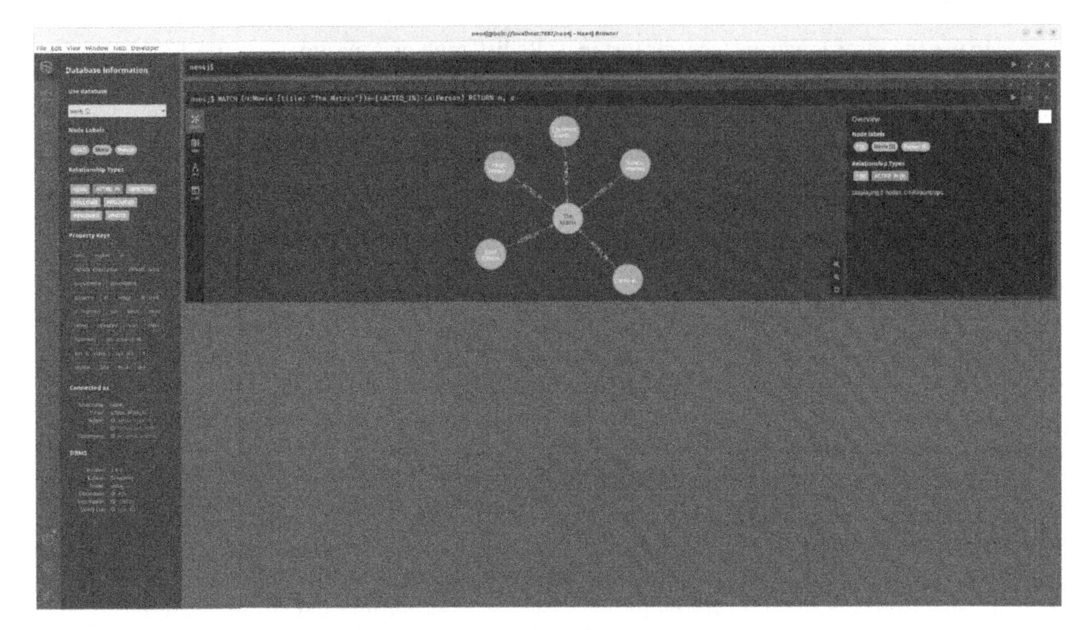

Figure 1.4 – Neo4j Browser

- **Neo4j Bloom**: A graph visualization application in which you can customize node styles (size, color, and so on) based on their labels and/or properties:

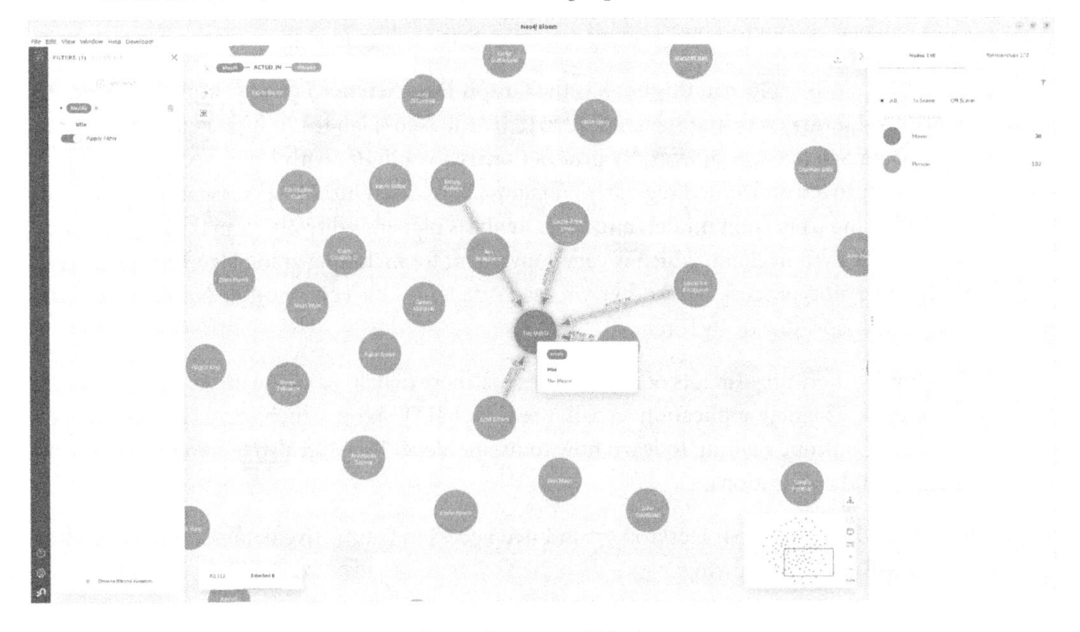

Figure 1.5 – Neo4j Bloom

- **Neodash**: This is a dashboard application that allows us to draw plots from the data stored in Neo4j, without having to extract this data into a DataFrame first. Plots can be organized into nice dashboards that can be shared with other users:

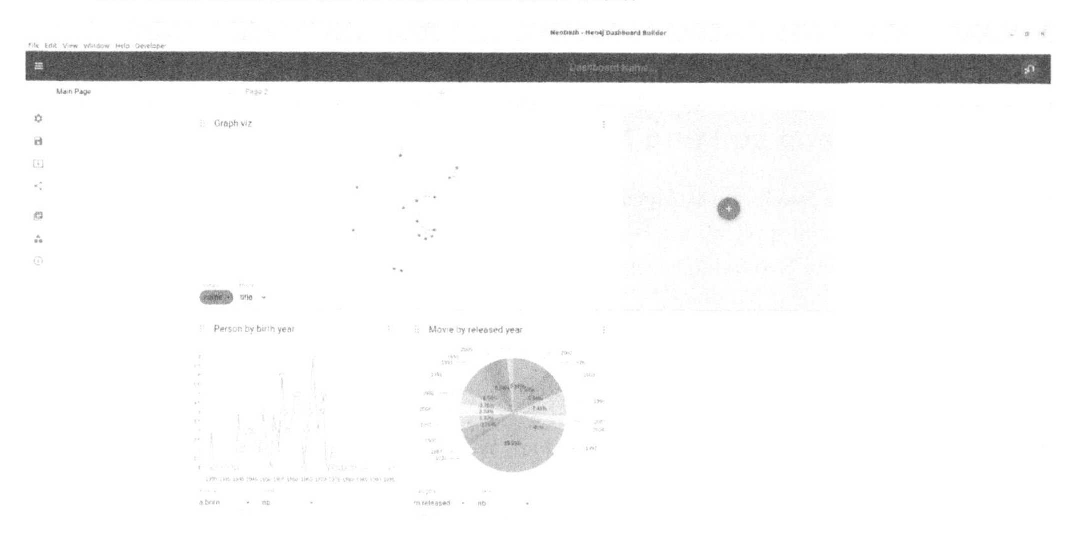

Figure 1.6 – Neodash

This list of applications is non-exhaustive. You can find out more here: `https://install.graphapp.io/`.

> **Good to know**
>
> You can create your own *graph application* to be run within Neo4j Desktop. This is why there are so many diverse applications, some of which are being developed by community members or Neo4j partners.

This section described Neo4j as a database and the various extensions that can be added to it to make it more powerful. Now, it is time to start using it. In the following section, you are going to install Neo4j locally on our computer so that you can run the code examples provided in this book (which you are highly encouraged to do!).

Setting up Neo4j

There are several ways to use Neo4j:

- Through short-lived time sandboxes in the cloud, which is perfect for experimenting
- Locally, with Neo4j Desktop
- Locally, with Neo4j binaries

- Locally, with Docker
- In the cloud, with Neo4j Aura (free plan available) or Neo4j AuraDS

For the scope of this book, we will use the Neo4j Desktop option, since this application takes care of many things for us and we do not want to go into server management at this stage.

Downloading and starting Neo4j Desktop

The easiest way to use Neo4j on your local computer when you are in the experimentation phase, is to use the Neo4j Desktop application, which is available on Windows, Mac, and Linux OS. This user interface lets you create Neo4j databases, which are organized into **Projects**, manage the installed plugins and applications, and update the DB configuration – among other things.

Installing it is super easy: go to the Neo4j download center and follow the instructions. We recap the steps here, with screenshots to guide you through the process:

1. Visit the Neo4j download center at `https://neo4j.com/download-center/`. At the time of writing, the website looks like this:

Figure 1.7 – Neo4j Download Center

2. Click the **Download Neo4j Desktop** button at the top of the page.
3. Fill in the form that's asking for some information about yourself (name, email, company, and so on).
4. Click **Download Desktop**.
5. Save the activation key that is displayed on the next page. It will look something like this (this one won't work, so don't copy it!):

```
eyJhbGciOiJQUzI1NiIsInR5cCI6IkpXVCJ9.
eyJlbWFpbCI6InN0ZWxsYTBvdWhAZ21haWwuY29tIiwibWl4cGFuZWxJZ
```

```
CI6Imdvb2dsZS1vYXV0a
...
...
```

The following steps depend on your operating system:

- On Windows, locate the installer, double-click on it, and follow the steps provided.
- On Mac, just click on the downloaded file.
- On Linux, you'll have to make the downloaded file executable before running it. More instructions will be provided next.

For Linux users, here is how to proceed:

6. When the download is over (this can take some time since the file is a few hundred MBs), open a Terminal and go to your download directory:

```
# update path depending on your system
$ cd Downloads/
```

7. Then, run the following command, which will extract the version and architecture name from the `AppImage` file you've just downloaded:

```
$ DESKTOP_VERSION=`ls -tr  neo4j-desktop*.AppImage | tail
-1 | grep -Po "(?<=neo4j-desktop-)[^AppImage]+"
$ echo ${DESKTOP_VERSION}
```

8. If the preceding `echo` command shows something like `1.4.11-x86_64.`, you're good to go. Alternatively, you can identify the pattern yourself and create the variable, like so:

```
$ DESKTOP_VERSION=1.4.11-x86_64.  # include the final dot
```

9. Then, you need to make the file executable with `chmod` and run the application:

```
# make file executable:
$ chmod +x neo4j-desktop-${DESKTOP_VERSION}AppImage
# run the application:
$ ./neo4j-desktop-${DESKTOP_VERSION}AppImage
```

The last command in the preceding code snippet starts the Neo4j Desktop application. The first time you run the application, it will ask you for the activation key you saved when downloading the executable. And that's it – the application will be running, which means we can start creating Neo4j databases and interact with them.

Creating our first Neo4j database

Creating a new database with Neo4j desktop is quite straightforward:

1. Start the Neo4j Desktop application.
2. Click on the **Add** button in the top-right corner of the screen.
3. Select **Local DBMS**.

   This process is illustrated in the following screenshot:

Figure 1.8 – Adding a new database with Neo4j Desktop

4. The next step is to choose a name, a password, and the version of your database.

> **Note**
>
> Save the password in a safe place; you'll need to provide it to drivers and applications when connecting to this database.

5. It is good practice to always choose the latest available version; Neo4j Desktop takes care of checking which version it is. The following screenshot shows this step:

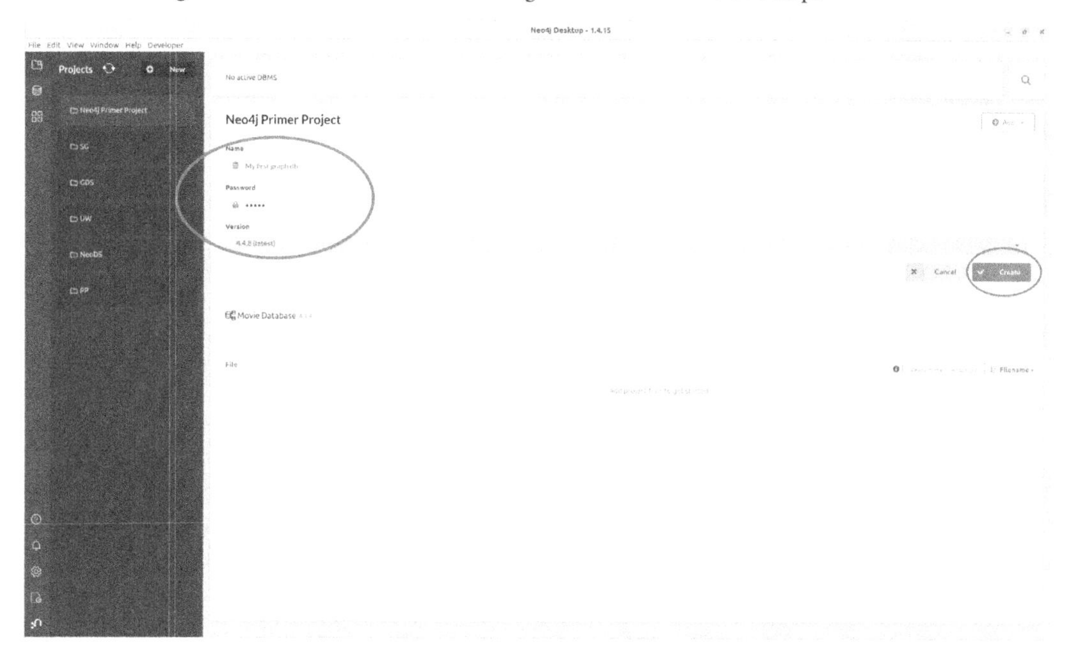

Figure 1.9 – Choosing a name, password, and version for your new database

6. Next, just click **Create**, and wait for the database to be created. If the latest Neo4j version needs to be downloaded, it can take some time, depending on your connection.
7. Finally, you can start your database by clicking on the **Start** button that appears when you hover your new database name, as shown in the following screenshot:

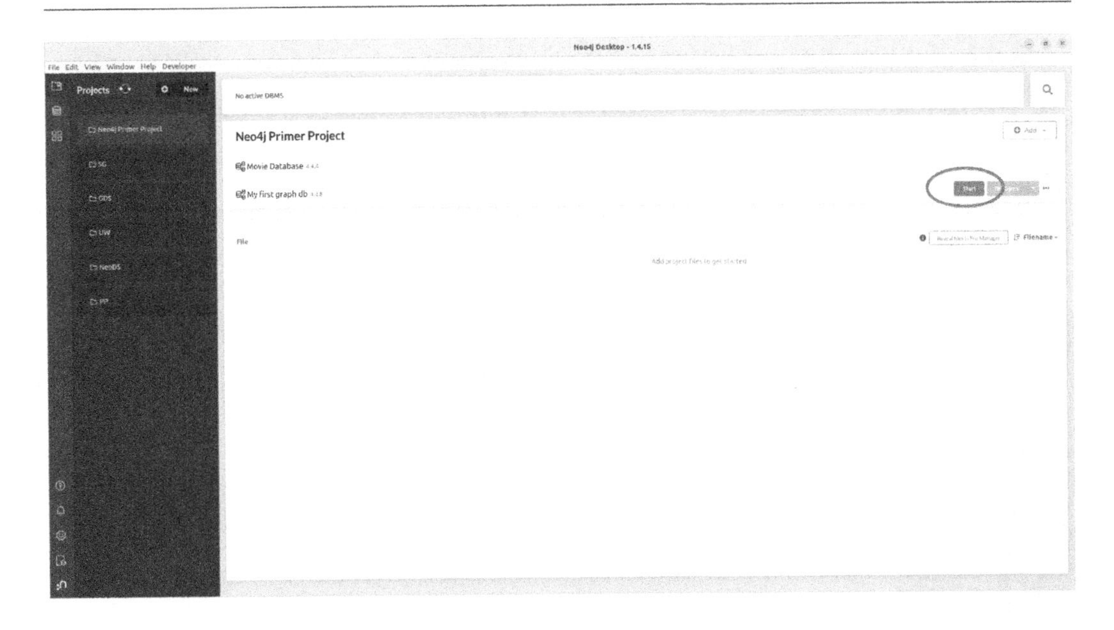

Figure 1.10 – Starting your newly created database

> **Note**
>
> You can't have two databases running at the same time. If you start a new database while another is still running, the previous one must be stopped before the new one can be started.

You now have Neo4j Desktop installed and a running instance of Neo4j on your local computer. At this point, you are ready to start playing with graph data. Before moving on, let me introduce Neo4j Aura, which is an alternative way to quickly get started with Neo4j.

Creating a database in the cloud – Neo4j Aura

Neo4j also has a **DB-as-a-service** component called *Aura*. It lets you create a Neo4j database hosted in the cloud (either on Google Cloud Platform or Amazon Web Services, your choice) and is fully managed – there's no need to worry about updates anymore. This service is entirely free up to a certain database size (50k nodes and 150k relationships), which makes it sufficient for experimenting with it. To create a database in Neo4j Aura, visit `https://neo4j.com/cloud/platform/aura-graph-database/`.

The following screenshot shows an example of a Neo4j database running in the cloud thanks to the Aura service:

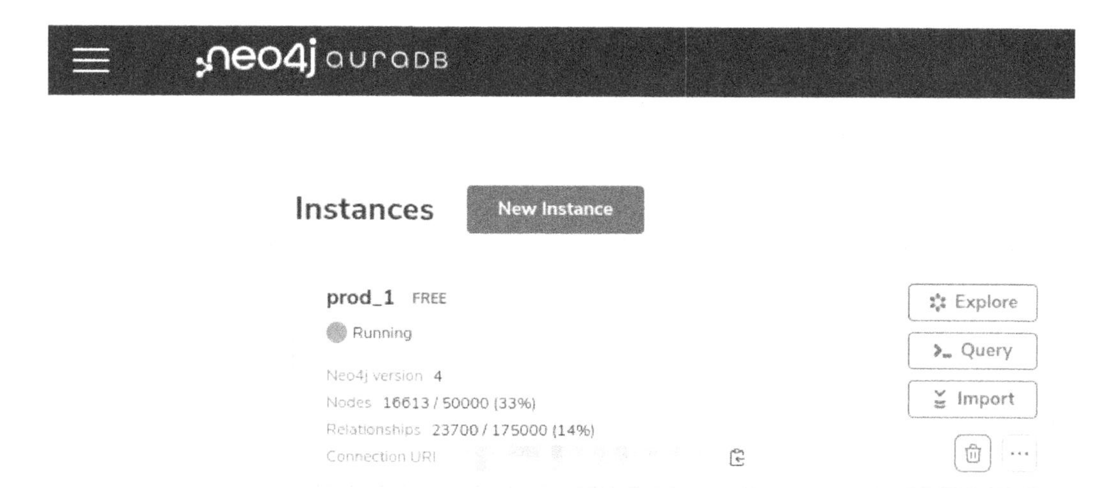

Figure 1.11 – Neo4j Aura dashboard with a free-tier instance

Clicking **Explore** opens Neo4j Bloom, which we will cover in *Chapter 3, Characterizing a Graph Dataset*, while clicking **Query** starts Neo4j Browser in a new tab. You'll be requested to enter the connection information for your database. The URL can be found in the previous screenshot – the username and password are the ones you set when creating the instance.

In the rest of this book, examples will be provided using a local database managed with the Neo4j Desktop application, but you are free to use whatever technique you prefer. However, note that some minor changes are to be expected if you choose something different, such as directory location or plugin installation method. In the latter case, always refer to the plugin or application documentation to find out the proper instructions.

Now that our first database is ready, it is time to insert some data into it. For this, we will use our first Cypher queries.

Inserting data into Neo4j with Cypher, the Neo4j query language

Cypher, as we discussed at the beginning of this chapter, is the query language developed by Neo4j. It is used by other graph database vendors, such as Redis Graph.

First, let's create some nodes in our newly created database.

To do so, open Neo4j Browser by clicking on the **Open** button next to your database and selecting **Neo4j Browser**:

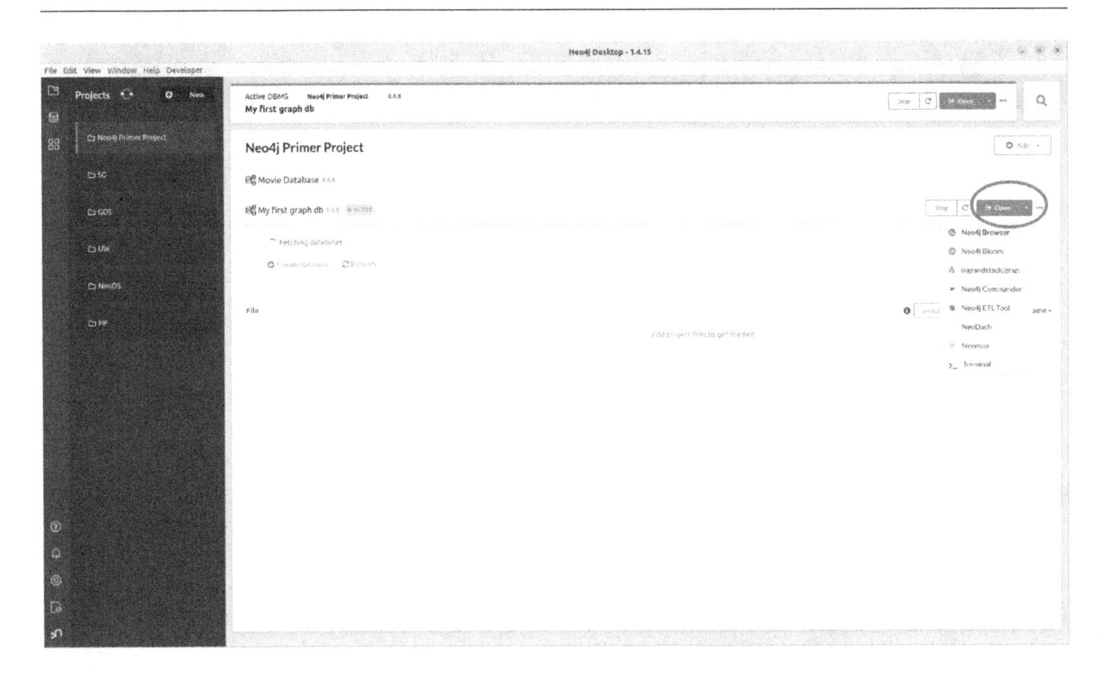

Figure 1.12 – Start the Neo4j Browser application from Neo4j Desktop

From there, you can start and write Cypher queries in the upper text area.

Let's start and create some nodes with the following Cypher query:

```
CREATE (:User {name: "Alice", birthPlace: "Paris"})
CREATE (:User {name: "Bob", birthPlace: "London"})
CREATE (:User {name: "Carol", birthPlace: "London"})
CREATE (:User {name: "Dave", birthPlace: "London"})
CREATE (:User {name: "Eve", birthPlace: "Rome"})
```

Before running the query, let me detail its syntax:

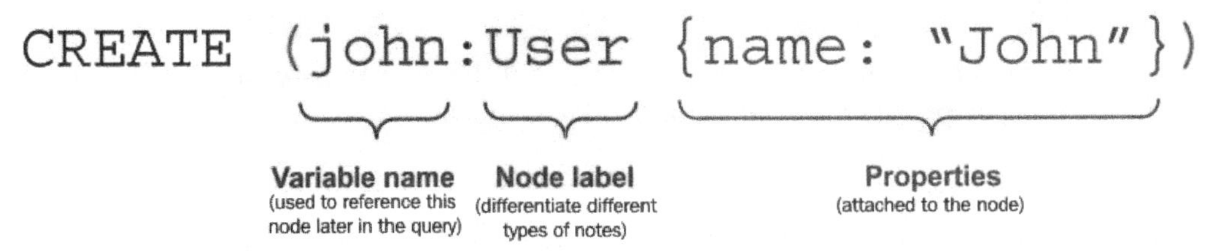

Figure 1.13 – Anatomy of a node creation Cypher statement

Note that all of these components except for the parentheses are optional. You can create a node with no label and no properties with `CREATE ()`, even if creating an empty record wouldn't be really useful for data storage purposes.

> **Tips**
>
> You can copy and paste the preceding query and execute it as-is; multiple line queries are allowed by default in Neo4j Browser.
>
> If the upper text area is not large enough, press the *Esc* key to maximize it.

Now that we've created some nodes and since we are dealing with a graph database, it is time to learn how to connect these nodes by creating edges, or, in Neo4j language, relationships.

The following code snippet starts by fetching the start and end nodes (`Alice` and `Bob`), then creates a relationship between them. The created relationship is of the `KNOWS` type and carries one property (the date Alice and Bob met):

```
MATCH (alice:User {name: "Alice"})
MATCH (bob:User {name: "Bob"})
CREATE (alice)-[:KNOWS {since: "2022-12-01"}]->(bob)
```

We could have also put all our `CREATE` statements into one big query, for instance, by adding aliases to the created nodes:

```
CREATE (alice:User {name: "Alice", birthPlace: "Paris"})
CREATE (bob:User {name: "Bob", birthPlace: "London"})
CREATE (alice)-[:KNOWS {since: "2022-12-01"}]->(bob)
```

> **Note**
>
> In Neo4j, relationships are directed, meaning you have to specify a direction when creating them, which we can do thanks to the > symbol. However, Cypher lets you select data regardless of the relationship's direction. We'll discuss this when appropriate in the subsequent chapters.

Inserting data into the database is one thing, but without the ability to query and retrieve this data, databases would be useless. In the next section, we are going to use Cypher's powerful pattern matching to read data from Neo4j.

Extracting data from Neo4j with Cypher pattern matching

So far, we have put some data in Neo4j and explored it with Neo4j Browser. But unsurprisingly, Cypher also lets you select and return data programmatically. This is what is called **pattern matching** in the context of graphs.

Let's analyze such a pattern:

```
MATCH (usr:User {birthPlace: "London"})
RETURN usr.name, usr.birthPlace
```

Here, we are selecting nodes with the `User` label while filtering for nodes with `birthPlace` equal to London. The `RETURN` statement asks Neo4j to only return the name and the `birthPlace` property of the matched nodes. The result of the preceding query, based on the data created earlier, is as follows:

"usr.name"	"usr.birthPlace"
"Bob"	"London"
"Carol"	"London"
"Dave"	"London"

This is a simple `MATCH` statement, but most of the time, you'll need to traverse the graph somehow to explore relationships. This is where Cypher is very convenient. You can write queries with an easy-to-remember syntax, close to the one you would use when drafting your query on a piece of paper. As an example, let's find the users known by Alice, and return their names:

```
MATCH (alice:User {name: "Alice"})-[:KNOWS]->(u:User)
RETURN u.name
```

The highlighted part in the preceding query is a graph traversal. From the node(s) matching label, `User`, and name, `Alice`, we are traversing the graph toward another node through a relationship of the `KNOWS` type. In our toy dataset, there is only one matching node, `Bob`, since `Alice` is connected to a single relationship of this type.

> **Note**
>
> In our example, we are using a single-node label and relationship type. You are encouraged to experiment by adding more data types. For instance, create some nodes with the `Product` label and relationships of the `SELLS/BUYS` type between users and products to build more complex queries.

Summary

In this chapter, you learned about the specificities of graph databases and started to learn about Neo4j and the tools around it. Now, you know a lot more about the Neo4j ecosystem, including plugins such as APOC and the **graph data science (GDS)** library and graph applications such as Neo4j Browser and Neodash. You installed Neo4j on your computer and created your first graph database. Finally, you created your first nodes and relationships and built your first Cypher `MATCH` statement to extract data from Neo4j.

At this point, you are ready for the next chapter, which will teach you how to import data from various data sources into Neo4j, using built-in tools and the common APOC library.

Further reading

If you want to explore the concepts described in this chapter in more detail, please refer to the following references:

- *Graph Databases, The Definitive Book of Graph Databases*, by I. Robinson, J. Webber, and E. Eifrem (O'Reilly). The authors, among which is Emil Eifrem, the CEO of Neo technologies, explain graph databases and graph data modeling, also covering the internal implementation. Very instructive!
- *Learning Neo4j 3.x - Second Edition*, by J. Baton and R. Van Bruggen. Even if written for an older version of Neo4j, most of the concepts it describes are still valid – the newer Neo4j versions have mostly added new features such as clustering for scalability, without breaking changes.
- *The openCypher project* (`https://opencypher.org/`) and *GQL specification* (`https://www.gqlstandards.org/`) to learn about graph query language beyond Cypher.

Exercises

To make sure you fully understand the content described in this chapter, you are encouraged to think about the following exercises before moving on:

1. Which information do you need to define a graph?
2. Do you need a graph dataset to start using a graph database?
3. True or false:
   A. Neo4j can only be started with Neo4j Desktop.
   B. The application to use to create dashboards from Neo4j data is Neo4j Browser.
   C. Graph data science is supported by default by Neo4j.
4. Are the following Cypher syntaxes valid, and why/why not? What are they doing?
   A. `MATCH (x:User) RETURN x.name`
   B. `MATCH (x:User) RETURN x`
   C. `MATCH (:User) RETURN x.name`
   D. `MATCH (x:User)-[k:KNOWS]->(y:User) RETURN x, k, y`
   E. `MATCH (x:User)-[:KNOWS]-(y) RETURN x, y`
5. Create more data (other node labels/relationship types) and queries.

2

Importing Data into Neo4j to Build a Knowledge Graph

As discussed in the previous chapter, you do not have to fetch a graph dataset specifically to work with a graph database. Many datasets we are used to working with as data scientists contain information about the relationships between entities, which can be used to model this dataset as a graph. In this chapter, we will discuss one such example: a Netflix catalog dataset. It contains movies and series available on the streaming platform, including some metadata, such as director or cast.

First, we are going to study this dataset, which is saved as a CSV file, and learn how to import CSV data into Neo4j. In a second exercise, we will use the same dataset, stored as a JSON file. Here, we will have to use the **Awesome Procedures on Cypher** (**APOC**) Neo4j plugin to be able to parse this data and import it into Neo4j.

Finally, we will learn about the existing public knowledge graphs, one of the biggest ones being Wikidata. We will use its public API to add more information to our graph database.

While exploring the datasets, we will also learn how to deal with temporal and spatial data types with Neo4j.

In this chapter, we're going to cover the following main topics:

- Importing CSV data into Neo4j with Cypher
- Introducing the APOC library to deal with JSON data
- Discovering the Wikidata public knowledge graph
- Enriching our graph with Wikidata information
- Importing data in the cloud

Technical requirements

To be able to reproduce the examples provided in this chapter, you'll need the following tools:

- Neo4j installed on your computer (see the installation instructions in the previous chapter).
- The Neo4j APOC plugin (the installation instructions will be provided later in this chapter, in the *Introducing the APOC library to deal with JSON data* section).
- Python and the Jupyter Notebook installed. We are not going to cover the installation instructions in this book. However, they are detailed in the code repository associated with this book (see the last bullet if you need such details).
- An internet connection to download the plugins and the datasets. This will also be required for you to use the public API in the last section of this chapter (*Enriching our graph with Wikidata information*).
- Any code listed in this book will be available in the associated GitHub repository, `https://github.com/PacktPublishing/Graph-Data-Science-with-Neo4j`, in the corresponding chapter folder.

Importing CSV data into Neo4j with Cypher

The **comma-separated values** (**CSV**) file format is the most widely used to share data among data scientists. According to the dataset of Kaggle datasets (`https://www.kaggle.com/datasets/morriswongch/kaggle-datasets`), this format represents more than 57% of all datasets in this repository, while JSON files account for less than 10%. It is popular for the following reasons:

- How it resembles the tabular data storage format (relational databases)
- Its closeness to the machine learning world of vectors and matrices
- Its readability – you usually just have to read column names to understand what it is about (of course, a more detailed description is required to understand how the data was collected, the unit of physical quantities, and so on) and there are no hidden fields (compared to JSON, where you can only have a key existing from the 1,000th record and later, which is hard to know without a proper description or advanced data analysis)

Knowing how to deal with such a data format is crucial to be able to use Neo4j in good conditions. In this section, we will introduce the CSV dataset we are going to use, try and extract entities (both nodes and relationships) we can model based on this data, and use Cypher to import the data into a new Neo4j graph.

Discovering the Netflix dataset

The dataset we are going to use in this section has been taken from `https://www.kaggle.com/datasets/shivamb/netflix-shows` and made available in this book's GitHub repository: `https://github.com/PacktPublishing/Graph-Data-Science-with-Neo4J/blob/main/Chapter02/data/netflix.zip`.

To understand its content, we are going to use the following notebook: `https://github.com/PacktPublishing/Graph-Data-Science-with-Neo4J/blob/main/Chapter02/notebooks/Netflix_dataset_analysis.ipynb`.

Let's discuss the code in the notebook:

1. Install the required packages if needed:

   ```
   !pip install pandas matplotlib seaborn
   ```

2. Import the required packages:

   ```
   import os
   import matplotlib.pyplot as plt
   import seaborn as sns
   import pandas as pd
   ```

3. Read data into a pandas DataFrame:

   ```
   # path relative to this notebook's location
   DATA_DIR = "../data"
   DATASET_FILE = "netflix.zip"

   data = pd.read_csv(os.path.join(DATA_DIR, DATASET_FILE))
   data.head(2)
   ```

4. List the columns with some details (name, number of non-null values, and the data type identified by pandas while loading the data):

   ```
   data.info()
   ```

5. The result of the preceding code is as follows:

   ```
   <class 'pandas.core.frame.DataFrame'>
   RangeIndex: 8807 entries, 0 to 8806
   Data columns (total 12 columns):
    #   Column          Non-Null Count  Dtype
   ```

```
---  ------          --------------  -----
 0   show_id         8807 non-null   object
 1   type            8807 non-null   object
 2   title           8807 non-null   object
 3   director        6173 non-null   object
 4   cast            7982 non-null   object
 5   country         7976 non-null   object
 6   date_added      8797 non-null   object
 7   release_year    8807 non-null   int64
 8   rating          8803 non-null   object
 9   duration        8804 non-null   object
 10  listed_in       8807 non-null   object
 11  description     8807 non-null   object
dtypes: int64(1), object(11)
memory usage: 825.8+ KB
```

6. Now, let's investigate the `type` column to learn more about the data that's being represented:

```
data.type.value_counts()
```

This yields the following output:

```
Movie      6131
TV Show    2676
Name: type, dtype: int64
```

This means our dataset contains information about more than 6,000 movies and more than 2,600 TV shows available on Netflix.

Let's go over some of the other columns:

- `title`: A string containing the title (in English)
- `director`: A string containing directors names
- `cast`: A string containing actors names
- `country`: A string containing countries the movie or show originates from
- `listed_in`: A string containing the movie's genres (or categories); for example, comedies, dramas, and so on
- `description`: A string containing the movie's description (in English)

7. Let's look more closely at the `cast` column by displaying the first 10 rows of this column:

```
0                                                       NaN
1      Ama Qamata, Khosi Ngema, Gail Mabalane, Thaban...
2      Sami Bouajila, Tracy Gotoas, Samuel Jouy, Nabi...
3                                                       NaN
4      Mayur More, Jitendra Kumar, Ranjan Raj, Alam K...
5      Kate Siegel, Zach Gilford, Hamish Linklater, H...
6      Vanessa Hudgens, Kimiko Glenn, James Marsden, ...
7      Kofi Ghanaba, Oyafunmike Ogunlano, Alexandra D...
8      Mel Giedroyc, Sue Perkins, Mary Berry, Paul Ho...
9      Melissa McCarthy, Chris O'Dowd, Kevin Kline, T...
Name: cast, dtype: object
```

Here, we can see the following:

- Some observations do not have cast information (rows with NaN values).
- For those with non-null data, the column contains a comma-separated list of all actors in the show. This makes sense since the relationship between a movie and the actors playing in it is not a one-to-one relationship (one movie, one actor) but a one-to-many relationship (one movie, many actors).

Now that we have a better understanding of the dataset's content, it is time to think about graphs and define the entities of importance to us and how we can model them in a graph database such as Neo4j.

Defining the graph schema

Thanks to the dataset exploration we performed in the previous section, we can identify a few entities described by our dataset:

- Movies
- TV shows
- Directors
- Actors
- Genres

We can create a graph schema similar to the one displayed in the following figure:

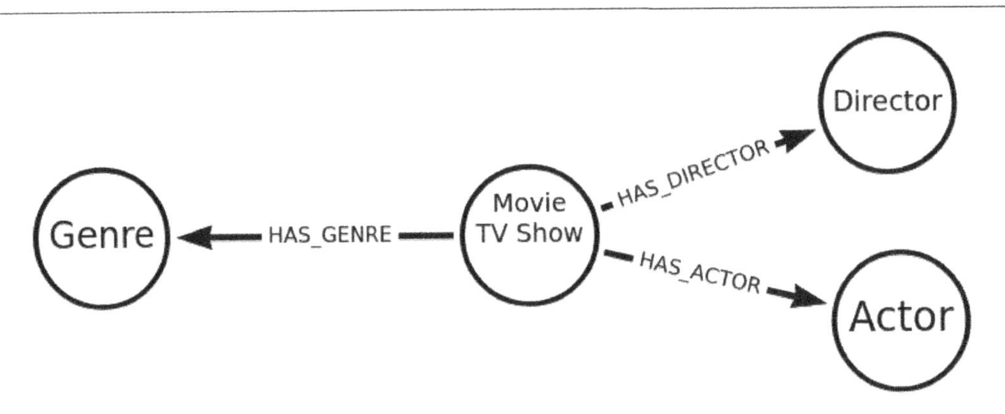

Figure 2.1 – Schema of the graph that can be created from the Netflix dataset

In this schema, movies and TV shows are grouped under the same label (that is, "collection"), while actors and directors are two separate collections. Relationships between nodes have obvious meanings: the tie between a movie and an entity, `X`, means `HAS_X`.

However, here, we can easily identify at least one issue: what happens to actors who are also directors (of the same movie or another one)? With this schema, they will be duplicated, and it will be more difficult to find out such patterns.

To overcome this issue, we propose the following schema instead:

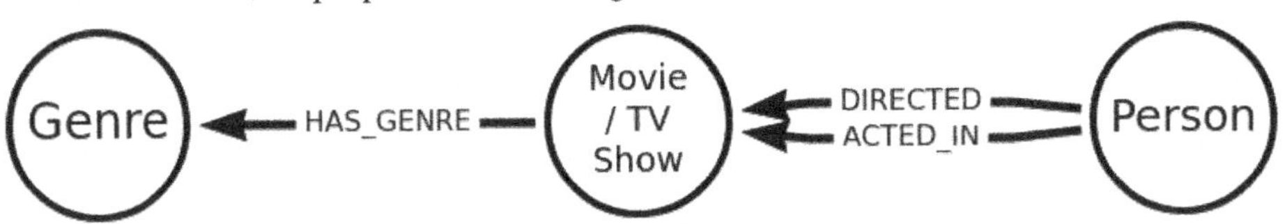

Figure 2.2 – Updated graph schema for the Netflix dataset

Here, both actors and directors are saved with the `Person` label. The way to remember their role in a given movie is saved through the `DIRECTED` or `ACTED_IN` relationship type.

Now, let's learn how to parse CSV files to import data into Neo4j.

Importing data

To import CSV data into Neo4j, we are going to use the `LOAD CSV` statement (`https://neo4j.com/docs/cypher-manual/current/clauses/load-csv/`). First, let's understand how to build such a statement.

LOAD CSV structure

A `LOAD CSV` statement looks like this, where optional parameters are denoted with square brackets:

```
LOAD CSV
[WITH HEADERS]
FROM 'file:///file_in_import_folder.csv'
AS line
[FIELDTERMINATOR ',']
// do stuffs with 'line'
```

Let's analyze this code block line by line:

1. First is the main statement, instructing Cypher we are going to read a CSV file.
2. On *line 2*, we can, if applicable, tell Cypher that our CSV file contains headers. If so, we will be able to access each field using its column's name; otherwise, only the column index is usable.
3. On *line 3*, we provide the path to the file to be imported. The file needs to be located in the import folder of your database. We'll learn how to find this folder after this explanation.

On *line 4*, we give each row of the file a name, which we will be able to use in the rest of the query to access fields. For instance, if `WITH HEADERS` is passed, we will be able to use `line.show_id`.

If your CSV file does not use the default comma separator, you can still parse it with Cypher using the optional `FIELDTERMINATOR` keyword.

> **Note**
>
> The line split here is done only for teaching purposes; Cypher is not indentation-sensitive. In the rest of this chapter, I will concatenate some of these lines to gain space.

Let's give it a try. First, we need to copy our data file into the Neo4j import folder. To do so, open Neo4j Desktop and create a new database, as we did in *Chapter 1, Introducing and Installing Neo4j*. You can start the database and follow these steps to locate your import folder:

1. Hover over the database's name. Locate the icon with the three dots (**...**) at the end of the line and click on this icon.
2. Click on **Open folder**.
3. Click on **Import**.

These steps can be visualized in the following screenshot:

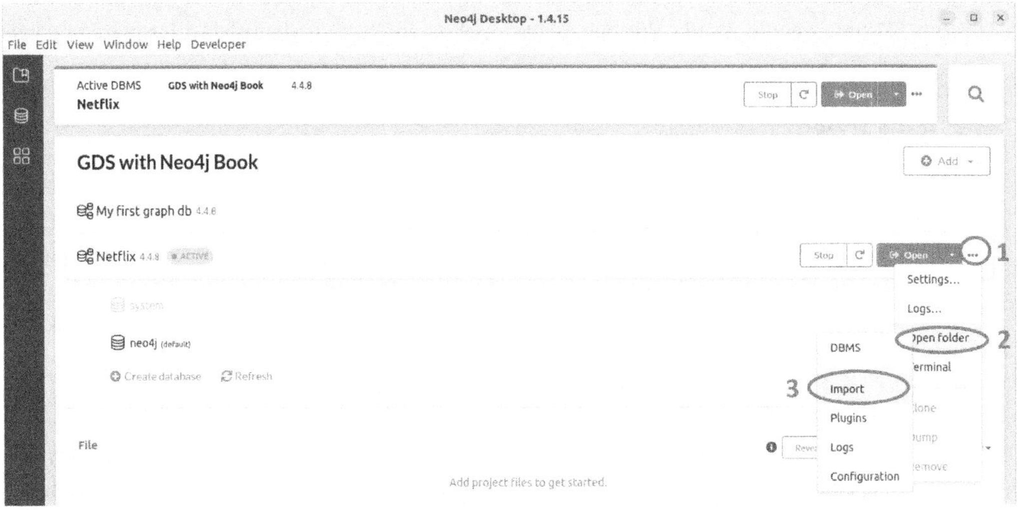

Figure 2.3 – How to open the Neo4j import folder with Neo4j Desktop

4. After performing this operation, the import folder will be opened in your file browser application. You can copy the file we studied earlier in this chapter and paste it into this directory.

> **Note**
>
> You'll have to unzip the file; at the time of writing, Cypher can't read compressed formats. The file is decompressed into `netflix/netflix_titles.csv`, which is the path we are going to use in the rest of this chapter. Update it accordingly if your path is different.

Finally, we can try and load some data into Neo4j. The beginning of the following code block is similar to the `LOAD CSV` structure we saw earlier with the proper path to our data file:

```
LOAD CSV WITH HEADERS
FROM 'file:///netflix/netflix_titles.csv' AS line
CREATE (:Movie {
    id: line.show_id,
    title: line.title,
    releaseYear: line.release_year
  }
)
```

The last statement is performing operations in the database and creating a node with the `Movie` label and three properties: `id`, `title`, and `releaseYear`.

Note about naming conventions

I am using the camel case notation of `releaseYear` instead of the snake case of `release_year` for properties stored in Neo4j to match Neo4j's conventions. This is not mandatory and you can name your properties with underscores in Neo4j if you prefer.

Executing the preceding query in Neo4j Browser yields the following result:

```
Added 8807 labels, created 8807 nodes, set 26421
properties, completed after 292 ms.
```

You can check the graph's content with this query:

```
MATCH (n) RETURN n LIMIT 20
```

As you can see, we have parsed the CSV file and imported all movies and TV shows it contains, storing the movie's ID, title, and release year as node properties. But the dataset contains more information, such as the countries of origin of the movies we are going to deal with now. To do so, we need to learn how to parse the comma-separated format of the dataset.

Parsing a comma-separated list in Cypher

The *country* column of our dataset contains a string, made up of a list of country names, separated by a comma – for instance, `"United States, India, France"`. Our goal is to be able to extract each of these names and create a single node for each.

Let's consider the following code:

```
WITH "United States, India, France" AS countries_as_string
WITH split(countries_as_string, ",") AS countries_as_list
UNWIND countries_as_list AS country_name
RETURN trim(country_name)
```

It is performing the following operations:

1. On the first line, we are creating a variable called `countries_as_string` with the format we want to parse.
2. On the second line, we are using the Cypher split function to extract a real list from this string. At this stage, `country_as_list` looks like this:

   ```
   ["United States", " India", " France"]
   ```

3. Then, we are using the `UNWIND` statement. This is where the list is iterated and each item is treated separately.
4. Finally, we are returning the trimmed country names to remove the extra remaining spaces.

   The result of the preceding query is as follows:

```
"United States"
"India"
"France"
```

   As you can see, our list was parsed and we ended up with one row per item in the list. Based on this, we can create one node for each of these items.

5. Going back to our `LOAD CSV` statement, we are going to use this new trick to parse directors for each movie, and create the associated `Person` nodes:

```
LOAD CSV WITH HEADERS
FROM 'file:///netflix/netflix_titles.csv' AS line
WITH split(line.director, ",") as directors_list
UNWIND directors_list AS director_name
CREATE (:Person {name: trim(director_name)})
```

6. If you run this query, you'll see a result similar to the following:

```
Added 6978 labels, created 6978 nodes, set 6978
properties, completed after 50 ms.
```

7. However, if you investigate the created graph, you'll notice a lot of duplicated persons. Indeed, for each director name in the `director` column, we have created a new `Person` node, independently of whether this person has already been created or not.

   What we want to do here is check if a `Person` node with a name property equal to `director_name` already exists:

   - If not, then create it
   - If yes, do nothing

This rule can be achieved using the Cypher `MERGE` statement, which we will investigate now.

The Cypher MERGE statement

First, let's delete all our *Person* nodes:

```
MATCH (p:Person) DELETE p
```

Then, we must replace the CREATE statement of our previous query with the MERGE keyword:

```
LOAD CSV WITH HEADERS
FROM 'file:///netflix/netflix_titles.csv' AS line
WITH split(line.director, ",") as directors_list
UNWIND directors_list AS director_name
MERGE (:Person {name: director_name})
```

The result for this query is as follows:

```
Added 5120 labels, created 5120 nodes, set 5120 properties,
completed after 3233 ms.
```

Compared to our previous try, we have created fewer nodes (`5,120` instead of `6,978`). The price to pay is a much higher execution time because, for each `director_name`, the database is queried for the existence check.

> **Information**
>
> This time difference can be considerably reduced by creating an index on a person's name property: `CREATE INDEX idx_person_name FOR (p:Person) ON p.name`.

Merging nodes with more than one property

The `MERGE` statement performs an exact match on all properties passed between the brackets. If you have multiple properties but only one needs to be used as a unique identifier, you can still use `MERGE` in conjunction with a `SET`; here's an example:

```
MERGE (p:Person {name: "Alice"})
ON CREATE SET p.surname = "Ecila"  // executed only the first
time the node is created
ON MATCH SET p.birthDate = "XXXX-XX-XX" // executed if the node
already exists
```

Our database now contains movies and all the people who directed such movies. But we have not stored the relationship between a movie and its directors yet, making our database unusable as-is. It is time to fix it!

Creating relationships

We are going to create relationships with two methods. First, we are going to continue importing data incrementally and read the CSV file a third time (after reading it once to import movies and another time to import people). Next, we'll see if we can perform all three operations (creating movies, directors, and their relationships) with a single query.

Reading the file again

When reading the file, we want to create a relationship between the existing `Movie` and `Person` nodes. To do so, we are going to `MATCH` these nodes, based on a unique identifier:

- `id` for `Movie`
- `name` for `Person`

Once these two nodes have been fetched and are in the query's scope, we can create the relationship between them with a `CREATE` statement:

```
LOAD CSV WITH HEADERS
FROM 'file:///netflix/netflix_titles.csv' AS line
MATCH (m:Movie {id: line.show_id})
WITH m, split(line.director, ",") as directors_list
UNWIND directors_list AS director_name
MATCH (p:Person {name: director_name})
CREATE (p)-[:DIRECTED]->(m)
```

The preceding query produces the following result:

```
Created 6978 relationships, completed after 23168 ms.
```

You'll notice that the number of created relationships matches the number of duplicated nodes we created when we first attempted to import people, so we have all our relationships set up.

Let's just check the content of the database for a single movie:

```
MATCH (m:Movie {title: "Ray"})<-[:DIRECTED]-(p:Person)
RETURN m, p
```

The resulting graph can be seen in the following diagram:

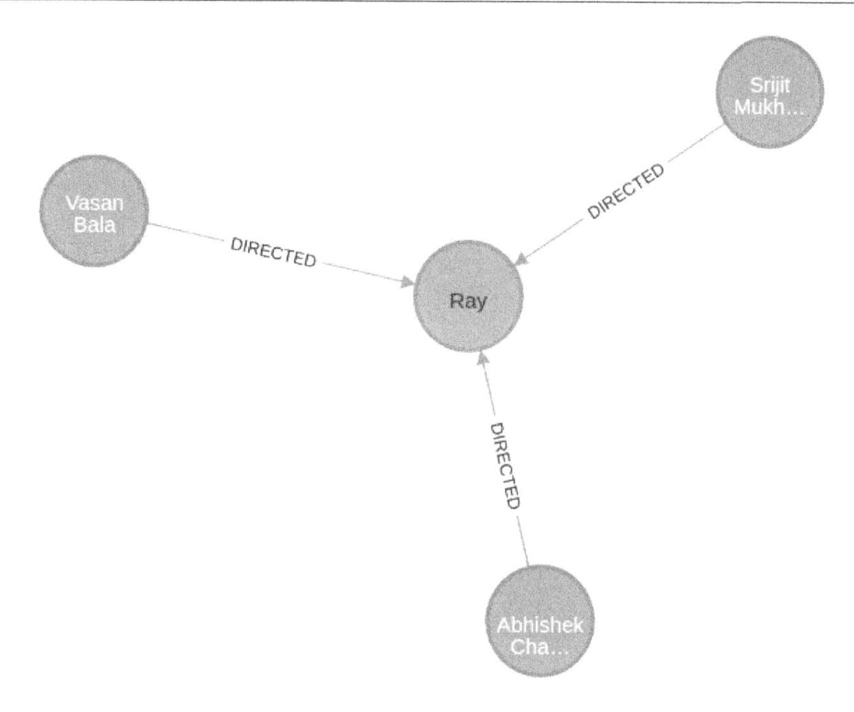

Figure 2.4 – Database content for the "Ray" movie, including its three directors

We have successfully imported data stored in a `CSV` file into Neo4j, including different node labels and a relationship. However, to do so, we read our file three times. In the following section, we are going to discuss how we can achieve the same result with a single query.

Creating nodes and relationships in one go

> **Warning**
>
> The query we are going to build next will recreate nodes and relationships we have already imported in the previous sections. If you want to try it, delete the imported data first with the `MATCH (n) DETACH DELETE n` Cypher query, which deletes all relationships and all nodes from the database.

Let me show you the code first; we will analyze it in the following bullet list:

```
LOAD CSV WITH HEADERS
FROM 'file:///netflix/netflix_titles.csv' AS line
CREATE (m:Movie {
    id: line.show_id,
    title: line.title,
```

```
    releaseYear: line.release_year
  }
)
WITH m, split(line.director, ",") as directors_list
UNWIND directors_list AS director_name
MERGE (p:Person {name: director_name})
MERGE (p)-[:DIRECTED]->(m)
```

This code can be explained as follows:

1. First, we create the movie node and give it an alias of `m`.
2. Then, we parse the director column, as we've already done, and `MERGE` the `Person` node. `MERGE` performs a `CREATE` or `MATCH` operation, meaning that the `p` node is always a `Person` node, either the existing node if such a node exists or a newly created one otherwise.
3. Finally, we can also use the `MERGE` keyword for relationships.

You can check the content of your database after this query. It is identical to the result we obtained with the previous method.

Practice!

The Netflix dataset contains other types of relationships, such as actors and countries. I'll let you import more data on your own. See the *Exercises* section at the end of this chapter for guidelines and answers.

Cypher's `LOAD CSV` statement is not the only way to load CSV data into Neo4j. Depending on the size of the dataset and the state of the database, you can also use the following tools:

- **Neo4j import tool**: This can only be used for a brand-new database that contains no data. It imports data blazingly fast, based on some configuration in the file headers or a separate file. More information about the import tool can be found here: `https://neo4j.com/docs/operations-manual/current/tutorial/neo4j-admin-import/`.
- **APOC**: The APOC plugin can also parse CSV files and proposes more options compared to the raw Cypher command. In particular, it can deal with quote characters and compressed files. See the full documentation here: `https://neo4j.com/labs/apoc/4.2/overview/apoc.load/apoc.load.csv/`.

Before moving on to the next section about JSON data, it is worth mentioning that Neo4j also provides an ETL tool to ease the process of translating relational databases to Neo4j. More information about this can be found here: `https://neo4j.com/developer/neo4j-etl/`.

This tour about CSV data is now over. It is time to switch to another popular data format: JSON.

Introducing the APOC library to deal with JSON data

The **JavaScript Object Notation** (**JSON**) file format is another data format you have probably used in your data science work. It is used by NoSQL document-like databases (or an equivalent of it, such as **Binary JSON** (**BSON**) for MongoDB). It is also one of the most used formats for data serialization and hence sharing data via web interfaces (APIs).

In this section, we will learn how to import JSON data into Neo4j. This format is not supported by Cypher directly, so we will have to rely on the APOC library to load such data. First, let's have a look at the dataset we are going to use in this section.

Browsing the dataset

The file we are going to use contains the same data we used in the previous section but in a different format. Here is an example record from the JSON file:

```
{'cast': [{'name': 'Billy Magnussen'},
            {'name': 'Ron Yuan'},
            {'name': 'Qu Jingjing'},
            {'name': 'Terry Chen'},
            {'name': 'Vanness Wu'},
            {'name': 'Jin Xing'},
            {'name': 'Philip Ng'},
            {'name': 'Xia Yu'},
            {'name': 'Yu Xia'}],
 'country': 'China, Canada, United States',
 'date_added': '2021-09-16',
 'director': [{'name': 'George Nolfi'}],
 'duration': '96 min',
 'listed_in': ['Action & Adventure', 'Dramas'],
 'release_year': 2017,
 'show_id': 's39',
 'title': 'Birth of the Dragon',
 'type': 'Movie'
}
```

As you can see, here are the changes that were made to the original CSV file:

- The `cast` and `director` fields have been transformed from a string containing a comma-separated list of items into a list of objects with a single key called `name`

- The `date_added` format was updated to match the ISO format: YYYY-MM-DD
- `listed_in` is now a true list of strings

The other fields have been kept unchanged. You may have noticed we are using three different formats to store a list of items: a string, as in the CSV example, a list of strings (`listed_in`), and a list of objects (`cast` and `director`). We will learn how to deal with each of these formats in the following sections. But first, let's discover the APOC library we've been talking about for a few pages!

Getting to know and installing the APOC plugin

The APOC plugin contains many different convenient procedures to help us with our day-to-day work with Neo4j. It provides features not covered by Cypher, which would require all of us to write sometimes complicated and often redundant code, using any of the Neo4j drivers in various languages if we had to perform such a task. And we have. Among other things, it exposes functions (returning a single value) and procedures (producing a result stream) related to the following:

- Graph refactoring (cloning nodes, changing a relationship's starting or ending node, and so on).
- Extensions of Cypher with, for instance, dynamic labels or property keys and periodic commits for all operations.
- Database introspection (graph schema, types of properties, and so on).
- Import from/export to files in different formats (JSON, XML, and so on).
- And much more! You are encouraged to take a look at the documentation, which is available here: `https://neo4j.com/labs/apoc/4.4/`.

In the rest of this chapter, we will only use the APOC capability to parse data in different formats, especially JSON.

Neo4j lets its users create custom functions or procedures. Implemented in Java and satisfying a predefined signature, users then just need to do the following:

1. Generate (or download) a JAR file.
2. Copy this file into the `plugins` directory of their Neo4j database.
3. Restart this database.

After that, all procedures defined in the JAR file can be called from Cypher using the `CALL` statement. We will use this statement extensively in this book, with APOC and then with the Graph Data Science Library, which follows the same implementation pattern.

However, if you are using Neo4j Desktop, as recommended in this book, installing APOC is even simpler than the steps described here.

Installing APOC with Neo4j Desktop

Follow these steps to create a new database and install APOC:

1. Start by creating a new database to hold this new dataset called **Netflix JSON**.
2. Click on your database's name.
3. Click on the **Plugins** tab.
4. Expand the **APOC** item.
5. Click **Install**.

The last three steps can be seen in the following screenshot:

Figure 2.5 – How to install the APOC plugin with Neo4j Desktop

That's it! You can now start your database, open Neo4j Browser, and start using the APOC procedures.

> **Tip**
>
> You can check the list of available procedures in your database by calling `SHOW procedures`. It displays the name of all procedures, their description, and their signature (name and type of input parameters and output variables).

Let's go back to our dataset and import our JSON data.

Loading data

To parse JSON data, we will use the following APOC procedure:

```
CALL apoc.load.json(URL) YIELD value
```

The *value* variable contains an object corresponding to each entry of the JSON array.

> **Note**
>
> The name after the `YIELD` statement must be one (or more) of the variables defined in the procedure signature. Using another name would result in an error like this:
>
> ```
> CALL apoc.load.json(URL) YIELD somethingElse
> // Unknown procedure output `somethingElse` (line 1, column 115
> (offset: 114))
> ```

To load our dataset, we will directly fetch the file from GitHub from the following URL:

```
https://raw.githubusercontent.com/PacktPublishing/Graph-Data-
Science-with-Neo4J/main/Chapter02/data/netflix.json
```

The syntax for reading and parsing JSON is as follows:

```
CALL apoc.load.json("https://raw.githubusercontent.com/
PacktPublishing/Graph-Data-Science-with-Neo4J/main/Chapter02/
data/netflix.json")
YIELD value
// do stuffs with 'value'
```

Note that if you have to load the file from a local folder, follow these steps instead:

1. Place the file in your import folder while following the procedure described in the previous section.
2. Update your database settings to allow APOC to read files there:

   ```
   apoc.import.file.enabled=true
   ```

3. Use the following path in the APOC procedure:

   ```
   file:///netflix.json
   ```

To simplify the notation for the following queries, I'll take advantage of a Neo4j Browser feature: **variables**.

Browser variables

Neo4j Browser can store variables that can be referred to in subsequent queries using the $ notation.

Inside the Browser's query box, type the following:

```
:param url=>"https://raw.githubusercontent.com/PacktPublishing/
Graph-Data-Science-with-Neo4J/main/Chapter02/data/netflix.json"
```

This defines a parameter called `url`. It can be accessed in the rest of your browser's session with the `$` notation; for instance:

```
RETURN $url
```

> **Note**
>
> If you are using a local path, update `url` accordingly.

Now, let's do stuff with each row. In the following four subsections, we are going to import the following:

- `Movies`
- `Countries`
- `Categories (listed_in)`
- `Persons` (actors and directors)

Before creating the nodes, let's create database indexes to make the operations faster.

Creating indexes

We are going to create indexes on properties that are used to uniquely identify nodes:

```
CREATE INDEX idx_movie_id FOR (m:Movie) ON m.id
CREATE INDEX idx_person_name FOR (p:Person) ON p.name
CREATE INDEX idx_country_name FOR (c:Country) ON c.name
CREATE INDEX idx_cat_name FOR (c:Category) ON c.name
```

With that, we can import data, starting with movies.

Creating movies

Exactly like we did for the CSV case, creating movies is as simple as this:

```
CALL apoc.load.json($url) YIELD value
CREATE (:Movie {
    id: value.show_id,
    title: value.title,
    releaseYear: value.release_year,
```

```
            dateAdded: value.date_added
        }
)
```

Creating countries is also similar since countries have the same format.

Creating countries

To create countries and link movies to their countries of origin, we can use the following query:

```
CALL apoc.load.json($url) YIELD value
MATCH (movie:Movie {id: value.show_id})
WITH movie, split(value.country, ",") as countries_list
UNWIND countries_list AS country_name
MERGE (country:Country {name: country_name})
MERGE (movie)-[:FROM]->(country)
```

Here, we have used the same trick as in the previous section to parse the country list string and extract individual values. This is because the country field was unchanged compared to the CSV case, and it was still a single string in our JSON file. Things are different for categories.

Creating categories

Categories are stored in an array of strings in our JSON file; here's an example:

```
'listed_in': ['Action & Adventure', 'Dramas'],
```

That means that APOC can parse this array properly, without us having to split a string. However, we still have to `UNWIND` categories to iterate over them:

```
CALL apoc.load.json($url) YIELD value
MATCH (movie:Movie {id: value.show_id})
UNWIND value.listed_in AS category
MERGE (cat:Category {name: category})
MERGE (movie)-[:LISTED_IN]->(cat)
```

This query creates one `Category` node for each category name encountered in the JSON file and creates relationships between a movie and each of the categories it belongs to according to Netflix.

Creating persons

Let's create `Person` nodes based on the `director` field, which is a list of objects with a single key called name:

```
'director': [{'name': 'George Nolfi'}]
```

Look at the following query:

```
CALL apoc.load.json($url) YIELD value
MATCH (movie:Movie {id: value.show_id})
UNWIND value.director AS director
MERGE (person:Person {name: director.name})
MERGE (person)-[:DIRECTED]->(movie)
```

The only difference compared to the previous query is that `director` is an object. We can access its attributes with the dot notation of `director.name`.

You can run the same query to import actors and create the `ACTED_IN` relationship between movies and their actors.

With that, you have created new nodes and relationships from data stored as JSON, setting their properties as required, such as the person's name, the movie's title, or the date it was added to Netflix. Let's stop a little bit and focus on this last field: `dateAdded`.

Dealing with temporal data

Date and time are special types of data. They can be compared and ordered so that we know which date comes first and can be subtracted to compute durations. A date can also be shifted by a certain time delta to find a new date. These operations are possible with Neo4j if the property is saved as a date type.

Currently, our `dateAdded` property is a string. This can be checked using the following query:

```
MATCH (m:Movie)
RETURN apoc.meta.cypher.type(m.dateAdded), count(*)
```

The result is as follows:

```
"apoc.meta.cypher.type(m.dateAdded)"|"count(*)"  |
|"STRING"                            |8797        |
|"NULL"                              |10          |
```

This means that none of the operations we talked about earlier would work. However, we can convert this string into a built-in date type using the `date()` function:

```
MATCH (m:Movie)
RETURN apoc.meta.cypher.type(date(m.dateAdded)), count(*)
```

This returns the following:

```
|"apoc.meta.cypher.type(date(m.dateAdded))"|"count(*)"|
|"LocalDate"                               |8797      |
|"NULL"                                    |10        |
```

Based on this, we can now perform operations such as finding the number of days since an item was first added to the platform and comparing it to the current date:

```
MATCH (m:Movie)
WHERE m.dateAdded is not null
WITH m,
        duration.inDays(
            date(m.dateAdded),
            date()  // date without parameter returns today's
date
        ).days as daysSinceAdded
RETURN m.title, m.dateAdded, daysSinceAdded
```

By sorting the results and returning only one row, we can get the first title ever added to Netflix:

```
MATCH (m:Movie)
WHERE m.dateAdded is not null
WITH m,
        duration.inDays(
            date(m.dateAdded),
            date()  // date without parameter returns today's
date
        ).days as daysSinceAdded
RETURN m.title, m.dateAdded
ORDER BY daysSinceAdded DESC
LIMIT 1
```

Time zones

Neo4j provides support for local dates and times with time zones. Refer to the documentation about `LocalDateTime`.

You can now import data into Neo4j from the two most popular data formats: CSV and JSON. APOC also provides utilities to import data from XML or GraphML, special markup languages used to store graph data (see *Chapter 1, Introducing and Installing Neo4j*), with dedicated procedures.

Using either CSV or JSON files, you have created your first knowledge graph about the Netflix catalog. In the following sections, you will learn how to take advantage of open knowledge graphs to add more information to your database.

Discovering the Wikidata public knowledge graph

Wikidata is a publicly available knowledge graph. It stores data in the RDF format. Like many RDF-like data stores, the query language used by Wikidata is SPARQL. Even if this is not the main topic of this book, we will see a couple of examples by the end of this chapter so that you can perform basic queries.

Wikidata data can be accessed via the following methods:

- A web browser, starting from the home page at `https://www.wikidata.org/`. Then, use the search bar to find the item of interest for you.
- A SPARQL playground, which is available at `https://query.wikidata.org/`.
- A public API using the endpoint: `http://query.wikidata.org/sparql?format=json&query=""`.

You are highly encouraged to navigate through Wikidata using your browser to understand the data format. You can, for instance, start from here:

- The Neo4j page at `https://wikidata.org/wiki/Q1628290`
- The page for India at `https://www.wikidata.org/wiki/Q668`
- The page dedicated to George Clooney (American actor): `https://www.wikidata.org/wiki/Q23844`

Data format

As quickly discussed in the previous chapter, RDF data is stored as triplets in the following format:

```
(SUBJECT, PREDICATE, OBJECT)
```

An example of an RDF triplet could be as follows:

```
(Neo4j, is an instance of, graph database)
(Neo4j, uses, Cypher Query Language)

(India, is an instance of, country)
(India, capital, New Delhi)
(India, coordinate location, 22°48'N, 83°0'E)
```

```
(George Clooney, is an instance of, human)
(George Clooney, country of citizenship, United States of
America)
(George Clooney, number of children, 4)
```

Each entity is identified by a key. For instance, Neo4j, as a graph database, is the `Q1628290` entity. Look again at the page URL given previously; you will easily understand how to find the ID of the other entities from their detailed web page.

Relationships or predicates also have description pages. Let's click on the **is an instance of** link in any of the pages we've seen so far. It will redirect us to `https://www.wikidata.org/wiki/Property:P31`, which contains information about what this predicate uses and the meaning of the relationship. Similarly, the country of citizenship predicate is described here: `https://www.wikidata.org/wiki/Property:P27`. You'll also notice that predicates have unique identifiers; `P31` is the ID of `is an instance of`, while `P27` identifies `country of citizenship`. We will use these IDs to follow relationships as we build queries in the next section.

Query language – SPARQL

SPARQL is a common language used to query data in RDF format. Let's build a query to extract information about a given person.

First, let's find this person by filtering entities by name:

```
SELECT ?person ?personLabel
WHERE {
  ?person rdfs:label "George Clooney"@en .

  SERVICE wikibase:label {bd:serviceParam wikibase:language
"en" .}

}
```

In the preceding SPARQL query, we are:

- Selecting (returning) the `?person` and `?personLabel` variables
- *person* must have the `"George Clooney"` label in English

The `SERVICE` line defines the service required to extract the `?personLabel` variable, instead of just `?person`, which, as you'll see if you run the preceding query, is just a raw identifier, not human-understandable.

Practice!

You can and are encouraged to copy and paste this query and the following ones into the Wikidata Query Service at `https://query.wikidata.org/`.

Starting from this `?person` node, similar to what we do with Cypher, we can traverse relationships. If you visit the United States of America page on Wikidata (`https://wikidata.org/wiki/Q30`), you'll be able to find the name and identifiers for such relationships (that is, predicates in RDF terms). The following table lists some of them:

Name	Identifier	Example Object
Instance of	P31	`"human"`
Country of citizenship	P27	`"United States of America"`
Date of birth	P569	`"6 May 1961"`
Native language	P103	`"English"`

Table 2.1 – Examples of instances in the Wikidata knowledge base

To find the country of citizenship of this person, we just have to follow the relationship, using the following syntax:

```
SELECT ?personLabel ?countryLabel
WHERE {
  ?person rdfs:label "George Clooney"@en ;
              wdt:P27 ?country .
  SERVICE wikibase:label {bd:serviceParam wikibase:language
"en" .}
}
```

By running this query in the Wikidata SPARQL playground, we obtain the following result:

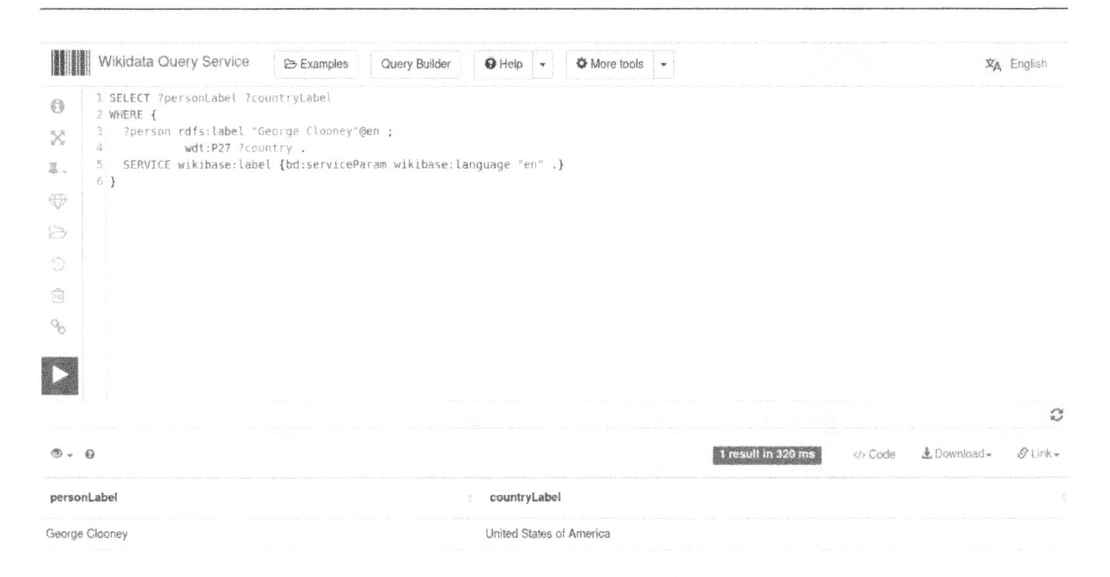

Figure 2.6 – SPARQL playground

Let's go ahead and make use of this information to extend our Netflix knowledge graph.

Enriching our graph with Wikidata information

In this section, we are going to use the preceding SPARQL query and the Wikidata query API to retrieve information about each person in our Neo4j graph and add their country of citizenship.

Loading data into Neo4j for one person

Using the previous query, we are going to query the Wikidata API SPARQL endpoint using APOC, and save the result into Neo4j:

1. Save the query as a parameter in Neo4j Browser:

```
:param query=>apoc.text.urlencode("SELECT ?personLabel
?countryLabel WHERE {?person rdfs:label 'George
Clooney'@en ; wdt:P27 ?country . SERVICE wikibase:label
{bd:serviceParam wikibase:language 'en' .}}")
```

2. Make sure you encode the query since it's going to be used in the query string to perform an `HTTP GET` query. You can see what the encoded query looks like by just using `RETURN $query`, which prints the following:

```
"SELECT+%3FpersonLabel+%3FcountryLabel+WHERE+%7B%3Fperson
+rdfs%3Alabel+%27George+Clooney%27%40en+%3B+wdt%3AP27+
%3Fcountry+.+SERVICE+wikibase%3Alabel+%7Bbd%3Aservice
Param+wikibase%3Alanguage+%27en%27+.%7D%7D"
```

3. Use the APOC procedure to load JSON from the API:

```
CALL apoc.load.json("http://query.wikidata.org/
sparql?format=json&query=" + $query)
YIELD value
RETURN value
```

The returned value is a JSON object. Look at its structure. The values of interest for us are located at the following path:

```
value.results.bindings[0].countryLabel.value
```

4. From there, we can update our existing node:

```
MATCH (person:Person {name: "George Clooney"})
CALL apoc.load.json("http://query.wikidata.org/
sparql?format=json&query=" + $query)
YIELD value
WITH person, value.results.bindings[0] as data
MERGE (country:Country {name: data.countryLabel.value})
CREATE (person)-[:IS_CITIZEN_OF]->(country)
```

5. Next, we can check the graph's content. In Neo4j Browser, you can click on the relationship's name from the left sidebar:

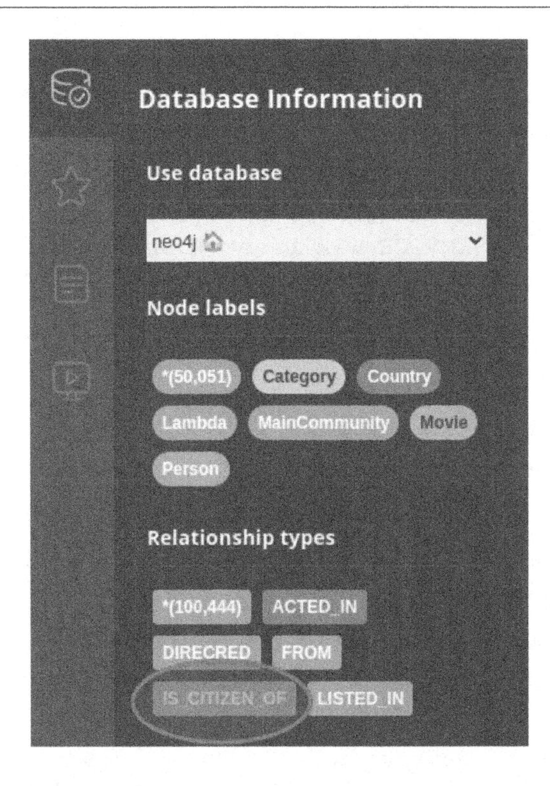

Figure 2.7 – Selecting a relationship type from Neo4j Browser's left sidebar

6. After clicking this name, the browser will create a new cell with a visualization of the relationship, as illustrated in the following screenshot:

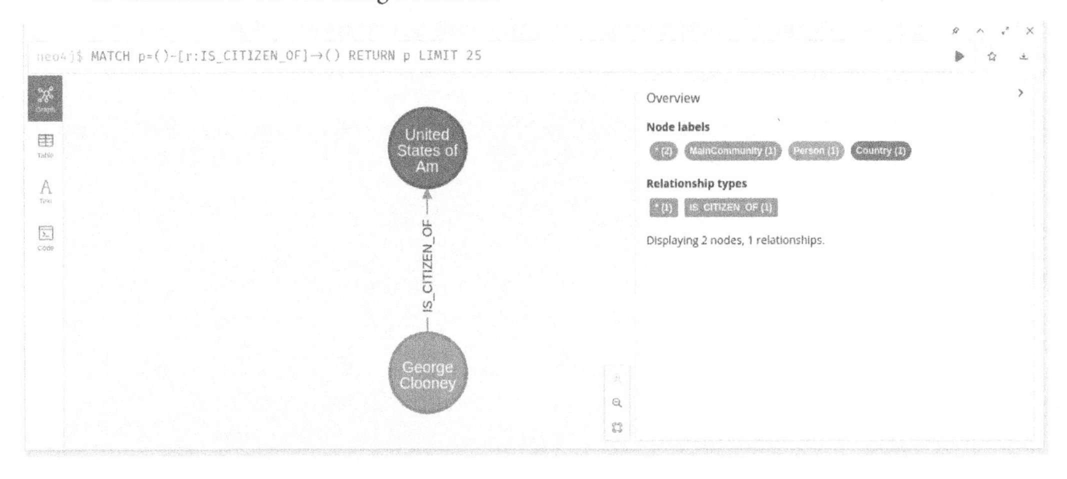

Figure 2.8 – Visualization of the newly created relationship, IS_CITIZEN_OF

With that, we have managed to import information from Wikidata into Neo4j for one person, namely George Clooney. In the next section, we are going to perform the same import for all existing people in our Neo4j graph.

Importing data for all people

To import the country of citizenship, we are going to dynamically build the SPARQL query for each node in our database. The following query does exactly that:

```
MATCH (p:Person)
WITH p, "SELECT DISTINCT ?countryLabel WHERE { ?person
rdfs:label \'" + p.name + "\'@en ; wdt:P27 ?country  .SERVICE
wikibase:label {bd:serviceParam wikibase:language \'en\'  .}}"
as query
RETURN p.name, query
LIMIT 10
```

The returned data looks like the following table:

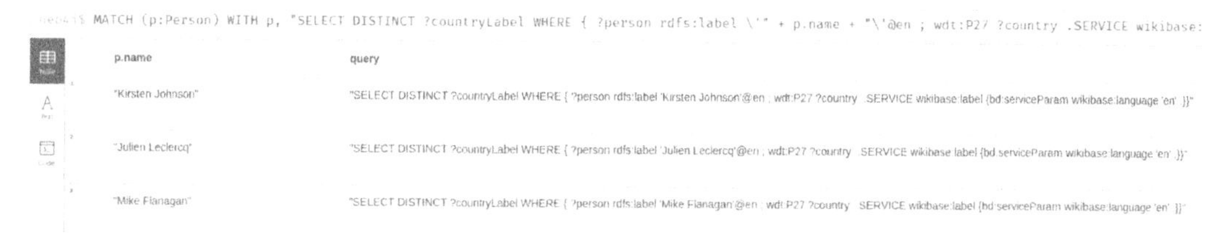

p.name	query
"Kirsten Johnson"	"SELECT DISTINCT ?countryLabel WHERE { ?person rdfs:label 'Kirsten Johnson'@en ; wdt:P27 ?country .SERVICE wikibase:label {bd:serviceParam wikibase:language 'en' .}}"
"Julien Leclercq"	"SELECT DISTINCT ?countryLabel WHERE { ?person rdfs:label 'Julien Leclercq'@en ; wdt:P27 ?country .SERVICE wikibase:label {bd:serviceParam wikibase:language 'en' .}}"
"Mike Flanagan"	"SELECT DISTINCT ?countryLabel WHERE { ?person rdfs:label 'Mike Flanagan'@en ; wdt:P27 ?country .SERVICE wikibase:label {bd:serviceParam wikibase:language 'en' .}}"

Figure 2.9 – Dynamically building a SPARQL query

Once we have the query, we can use the `apoc.text.urlencode` and `apoc.load.json` procedures as we did previously to extract the results for each node. However, this query has one major drawback: if any of the SPARQL queries fail, for whatever reason, then the whole Cypher query will fail, and no relationship will be created. To overcome this issue, we will use another APOC procedure, called `apoc.periodic.iterate`, in the following way:

```
CALL apoc.periodic.iterate(
  "MATCH (p:Person) WHERE NOT exists((p)-[:IS_CITIZEN_OF]-
>(:Country)) RETURN p LIMIT 500",
  'WITH p, "SELECT DISTINCT ?countryLabel WHERE { ?person
rdfs:label \'" + p.name + "\'@en ; wdt:P27 ?country  .SERVICE
wikibase:label {bd:serviceParam wikibase:language \'en\'
.}}" as query CALL apoc.load.json("http://query.wikidata.org/
sparql?format=json&query=" + apoc.text.urlencode(query)) YIELD
value WITH p, value.results.bindings as results UNWIND results
```

```
as r MERGE (c:Country {name: r.countryLabel.value}) MERGE (p)-
[:IS_CITIZEN_OF]->(c)',
      {batchSize: 100}
)
```

This looks quite complicated but let's explain it. The procedure accepts three parameters:

- The first parameter is a Cypher query returning a list of all the items we want to iterate on. The returned objects (p, in our case) are passed to the query in the second argument.

  Note that we are filtering the `Person` nodes that do not have any `IS_CITIZEN_OF` relationship yet so that we can restart the process if needed – for instance, if we reach the rate limit of the API so that it won't query the people for which we already have data.

- The second query uses the `p` variable and performs the operation we are interested in – in our case, building the SPARQL query, fetching data from Wikidata, and creating the relationships. The most difficult part is to escape the quotes properly with \ when we are inside the query.

- Finally, the last parameter is a configuration map, from which we only use the `batchSize` parameter. It means that for every 100 rows, the results will be saved to the database. If any of the batches fail, it won't affect the other ones.

> **Information**
>
> In the following chapters, we will need the citizenship information for all the people in the database. A full dataset will be provided so that you don't have to run this query for the full database; it takes some time to complete.

In this section, we used some shortcuts. For instance, we haven't considered the problem of homonyms.

We've only considered the first row of the result returned by Wikidata, assuming only one person matches a given name. As you can imagine, this is not true in a general sense. We could refine the query to make sure we are only retrieving actors and movie directors, for instance. You can do this as an exercise; it will be explained at the end of this book.

> **Note**
>
> In this section, we used built-in Neo4j tools and APOC, which we used earlier in this chapter, for simplicity. If you have to deal with RDF-like data such as the one from Wikidata extensively, be aware that there is a specific Neo4j plugin for that: *neosemantics*. See, for instance, the related page on the documentation, which demonstrates how to import Wikidata information into Neo4j: `https://neo4j.com/labs/neosemantics/how-to-guide/`.

Now, you can import data from CSV files, JSON files, or API and RDF databases. Before closing this dense chapter, we are going to address another special data type that's also supported in Neo4j: spatial data.

Dealing with spatial data in Neo4j

Neo4j has a built-in type for dealing with spatial data, but only for points so far (lines and polygons are not supported (yet)).

Wikidata contains coordinates information for many entities. For instance, each country has a location. We can extract it using the following query:

```
SELECT ?country ?countryLabel ?lat ?lon
WHERE {
  ?country rdfs:label "India"@en;
                wdt:P31 wd:Q6256.
     ?country p:P625 ?coordinate.
     ?coordinate psv:P625 ?coordinate_node.
     ?coordinate_node wikibase:geoLongitude ?lon.
     ?coordinate_node wikibase:geoLatitude ?lat.

  SERVICE wikibase:label {bd:serviceParam wikibase:language
"en" .}
}
```

> **Going further**
>
> Explanations of the new parts of this query are available here: `https://en.wikibooks.org/wiki/SPARQL/WIKIDATA_Precision,_Units_and_Coordinates#Coordinates`.

Let's replace our `$query` variable with the new one:

```
:param query=>apoc.text.urlencode("SELECT ?country
?countryLabel ?lat ?lon WHERE { ?country rdfs:label
'India'@en; wdt:P31 wd:Q6256. ?country p:P625 ?coordinate.
?coordinate psv:P625 ?coordinate_node. ?coordinate_
node wikibase:geoLongitude ?lon. ?coordinate_node
wikibase:geoLatitude ?lat. SERVICE wikibase:label
{bd:serviceParam wikibase:language 'en' .}}")
```

We can add a new property to the `Country` node:

```
MATCH (country:Country {name: "India"})
CALL apoc.load.json("http://query.wikidata.org/
sparql?format=json&query=" + $query)
YIELD value
WITH country, value.results.bindings[0] as data
SET country.point = point({latitude: toFloat(data.lat.value),
longitude: toFloat(data.lon.value)})
```

The advantage of this data type is that it allows us to perform spatial queries, especially based on distance. For instance, we can query the distance between India (the central point) and a user-defined location with this code:

```
MATCH (india:Country {name: "India"})
RETURN point.distance(
              india.point, // India center
              point({latitude: 45, longitude: 2}) // a new
point, somewhere in France
)
```

This overview of spatial data types in Neo4j closes this long chapter in which you have learned how to feed data into Neo4j from different data sources and deal with different data types.

In the next and last section of this chapter, we will discuss importing data into Neo4j when using the cloud version of Neo4j called Aura.

Importing data in the cloud

To import data into Neo4j Aura, the cloud-hosted Neo4j database, we can use the aforementioned method of reading files from an accessible URL. But Neo4j also provides a frontend application that can deal with CSV files only.

Starting from a Neo4j Aura console, as illustrated in *Figure 1.11* in *Chapter 1, Introducing and Installing Neo4j*, you can click on the **Import** button. That will open the data importer login window, as shown in the following screenshot:

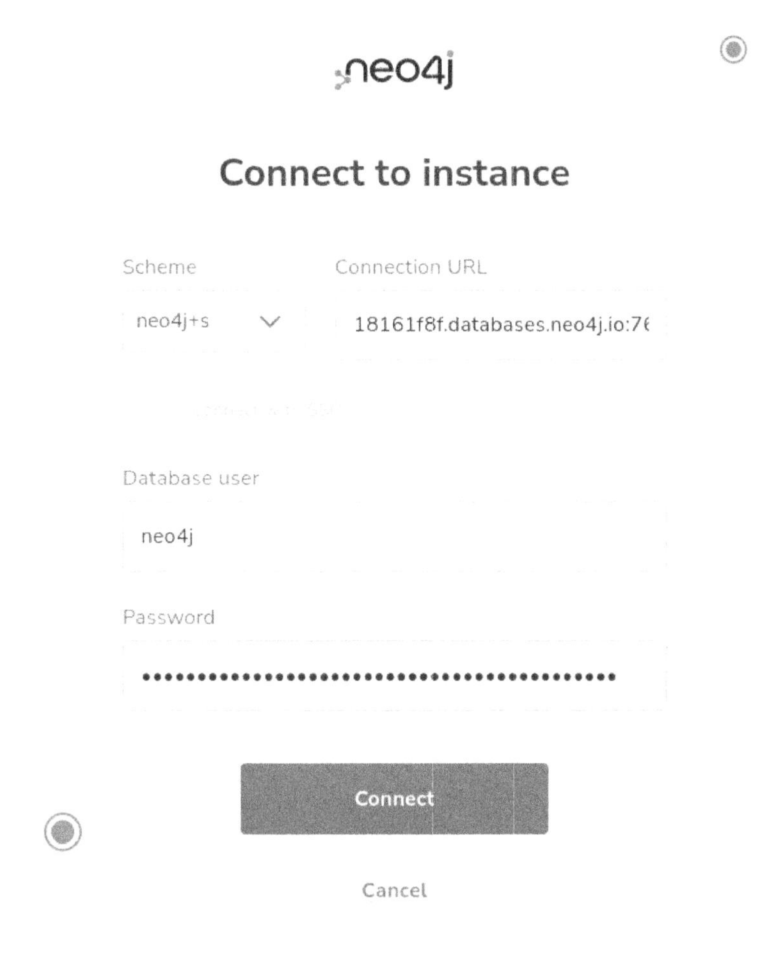

Figure 2.10 – Logging into Neo4j Data Importer

After entering your credentials, you will be redirected to the application. The UI is made up of three parts:

- The **Files** manager, on the left: The area where you can drag and drop the files to be imported
- The graph view (middle panel): This is where you can draw your graph schema, including nodes and relationships
- **Mapping details**: In this section, you can define node and relationship properties, and you can map them to the fields in the CSV files previously added to the first component

This UI is illustrated in the following screenshot:

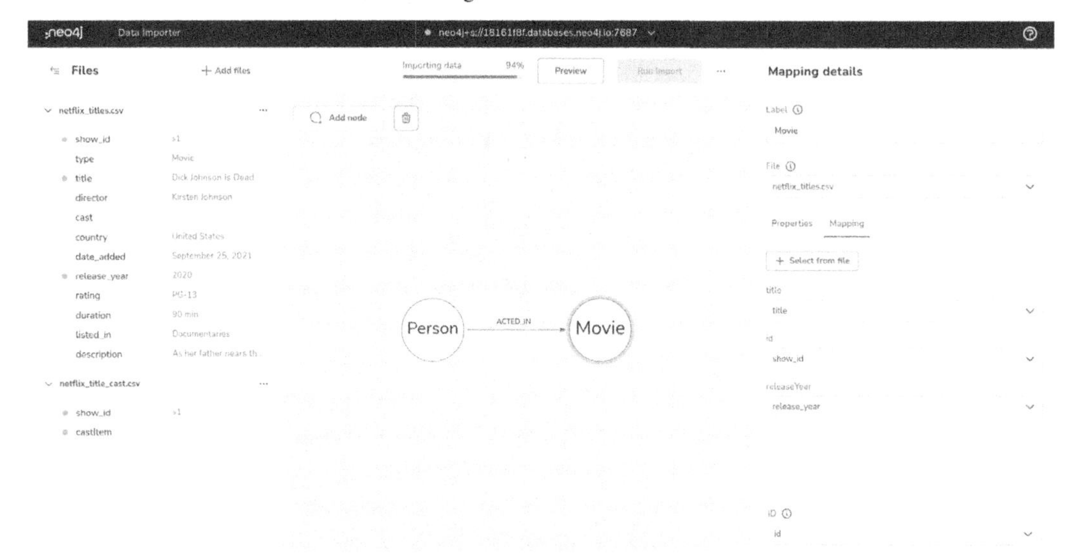

Figure 2.11 – Neo4j Data Importer

Let's detail the steps to import the Netflix dataset with this importer:

1. In terms of data preparation, the importer can't parse complex fields such as lists, so we have created an intermediary file called `netflix_title_cast.csv` containing two columns: `show_id` and a single actor name (`castItem`).
2. Then, both `netflix_title_cast.csv` and `netflix_title.csv` are added to the **Files** section.
3. Now, can add the `Person` node and draw a relationship from this node to another node that we will rename *Movie*.
4. In the **Properties** tab in the right panel, we must add properties for nodes:
   - `name` (string) for `Person`
   - `id` (string), `title` (string), and `releaseYear` (integer) for `Movie`
5. In the **Mapping** tab, we can map the node properties to the files headers:
   - Movie `title` property => the `title` column in the `netflix_title.csv` file
   - Movie `id` property => the `show_id` column in the `netflix_title.csv` file
   - We do the same for the other properties, including the `Person` properties

6. We can also define, at the bottom of the right panel, the unique identifier property for each node: `id` for movies and `name` for people.
7. Then, we can define how to create relationships between nodes. This configuration is illustrated in the following screenshot:

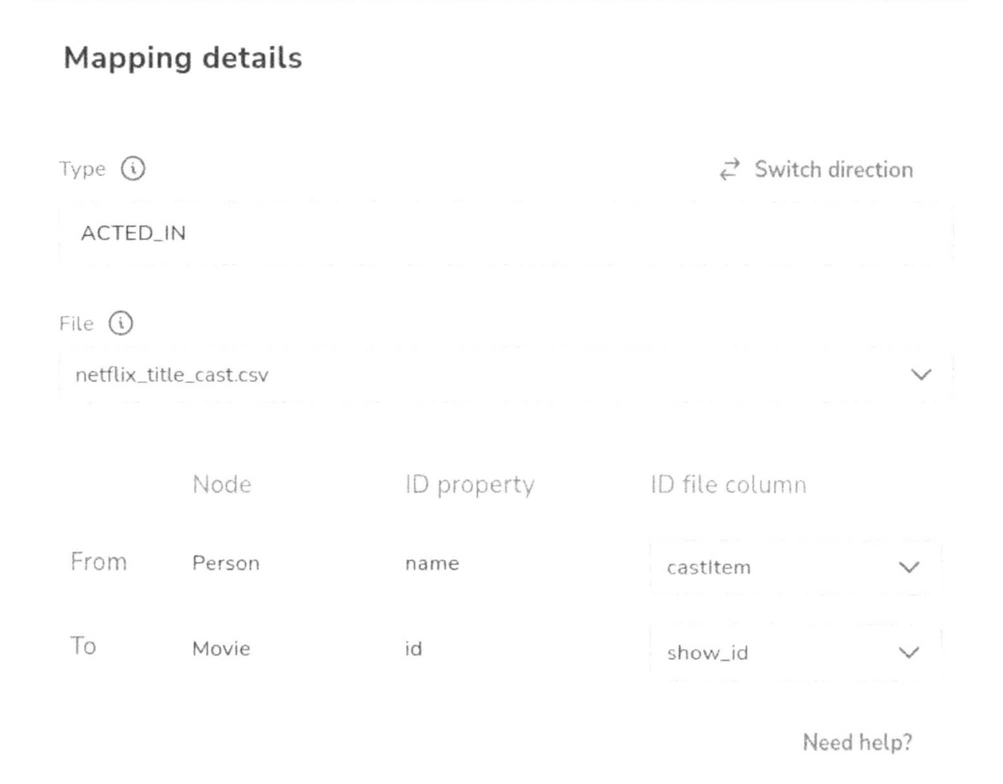

Figure 2.12 – Relationship mapping configuration

8. Finally, we can click the **Run Import** button at the top of the screen, which will process the data to be imported into Neo4j.

   With our dataset, the import produces the following results:

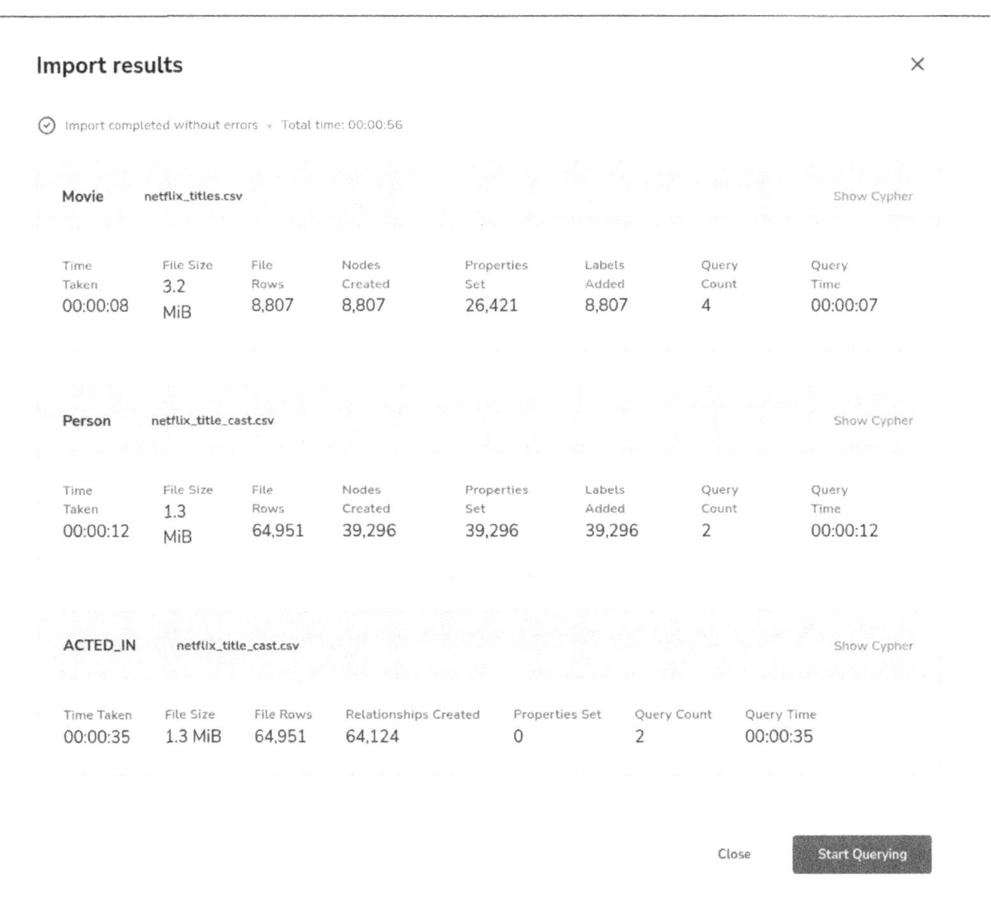

Figure 2.13 – Import result

This result lists some statistics about the import process, especially the number of created nodes and the relationships of each type.

Once the data has been imported, you can proceed to **Start Querying** to open Neo4j Browser and start working with your newly imported data.

Summary

Importing existing data into a brand-new database is always a concern, as we covered in this chapter. From a flat (a non-graph format) file, you can identify node labels and relationship types between them, transforming a flat dataset into a real graph. Whether your data is stored as CSV, JSON, on your local disk or distant server, or via an API endpoint, you can now load this data into Neo4j and start exploring your graph. You also learned about the Neo4 Data Importer tool, which is used to import data stored as CSV files in a cloud-hosted Neo4j database (Aura).

You also learned about public knowledge graphs, such as Wikidata, which can be used to extend your knowledge by importing more data about a specific topic.

Finally, you learned how to import your data into the cloud thanks to the Neo4j Data Importer application.

Being able to create a graph dataset is only the beginning, though. Like any dataset, graph datasets are very different from one to another. While in a tabular dataset, you would analyze the type of columns and the distribution of variables, as well as compute some statistical indicators such as the mean or standard deviation, graphs also have key metrics to help you get to know them better.

In the next chapter, we will start exploring such metrics while introducing the Neo4j Python driver, which allows you to extract data stored in Neo4j from your Python code.

Further reading

The following resources might help you gain a better understanding of the topics covered in this chapter:

- Graph data modeling is covered in *The Practitioner's Guide to Graph Data*, by D. Gosnell and M. Broecheler (O'Reilly)
- The SPARQL language to model RDF data is described in detail in *Semantic Web for the Working Ontologist*, by D. Allemang and J. Hendler (Morgan Kaufmann)

Exercises

To make sure you understand the topics covered in this chapter before moving on to the next one, you are encouraged to think about the following:

1. What is the advantage of a `MERGE` statement over `CREATE`?
2. Can raw Cypher parse JSON data? What tool should you use for that?
3. Practice! Using the Netflix dataset, set the movie's genres contained in the `listed_in` column in the CSV dataset (assume `Movies` has already been imported).
4. Practice! Using the Netflix JSON dataset, write a Cypher query to import actors (assume `Movies` has already been imported).
5. Knowing that a given user – let's call her Alice – watched the movie named *Confessions of an Invisible Girl*, what other Netflix content can we recommend to Alice?
6. Practice! Refine the SPARQL query we've built to make sure the person is an actor or movie director.

   Help: You can use *Robert Cullen* as an example.

The answers are provided at the end of this book.

Part 2 – Exploring and Characterizing Graph Data with Neo4j

In this part, you will learn how to characterize a graph dataset: which information and metrics can be provided to distinguish one dataset from another. You will be introduced to the Neo4j **Graph Data Science** (**GDS**) library and the most common graph algorithms to analyze the graph topology, finding important nodes or clusters of highly connected nodes. You will also discover some tools to visually explore graph data.

This part includes the following chapters:

3

Characterizing a Graph Dataset

Two graphs can differ in many ways, depending on their number of nodes or types of edges, for instance. But many more metrics exist to characterize them so that we can get an idea of the graph based on some numbers. Just as the mean value and standard deviation help in comprehending a numeric variable distribution, graph metrics help in understanding the graph topology: is it a highly connected graph? Are there isolated nodes?

In this chapter, we are going to learn about a few metrics for characterizing a graph. Focusing on the degree and degree distribution, this will be an opportunity for us to draw our first plot using the **NeoDash** graph application. We will also use the Neo4j Python driver to extract data from Neo4j into a DataFrame and perform some basic analysis of this data.

In this chapter, we're going to cover the following main topics:

- Characterizing a graph from its node and edge properties
- Computing the graph degree distribution
- Installing and using the Neo4j Python driver
- Learning about other characterizing metrics

Technical requirements

To be able to reproduce the examples provided in this chapter, you'll need the following tools:

- Neo4j installed on your computer (see the installation instructions in *Chapter 1, Introducing and Installing Neo4j*).
- The necessary Python and Jupyter notebooks installed. We are not going to cover the installation instructions in this book.

- You'll also need the following Python packages:
  - `matplotlib`
  - `pandas`
  - `neo4j`
- An internet connection to download the plugins and the dataset and to use the public API in the last section of this chapter.
- Any code listed in the book will be available in the associated GitHub repository, `https://github.com/PacktPublishing/Graph-Data-Science-with-Neo4j`, in the corresponding chapter folder.

Characterizing a graph from its node and edge properties

There is not a single type of graph. Each of them has specific characteristics, depending on the modeled process. This section describes some of the characteristics of a graph you should question when starting your journey with a new dataset.

Link direction

Links between nodes can be directed (and are then called **arcs** *in graph theory*) or undirected (and are called **edges**).

While graph theory makes the distinction between directed and undirected links in their naming, the graph database vocabulary usually doesn't, and all links are called **edges** or **relationships**, regardless of whether they're considered directed or not. In a more general way, I'll stick to the wording used within the Neo4j Graph Data Science Library, which may sound inaccurate to graph theorists.

Undirected graphs include the following:

- **Facebook social network**: If you are connected to `X`, `X` is also connected to you.
- **Co-authorship graphs**: Such graphs connect researchers if they have co-authored the same paper. By definition, the relationships are symmetric and undirected.

On the other hand, here are some examples of directed graphs:

- **Twitter network**: You can follow someone who is not following you and vice versa
- **The internet linking**: If we create a graph where nodes are web pages and edges are the web links between them, this graph will be directed (a page points toward another one, but the contrary is usually not true)
- **Road network**: A road network where links represent streets and nodes intersection is a directed network that takes into account traffic direction restrictions on the different road segments

> **Note**
>
> Edge orientation is one key element to take into account when dealing with graphs. It has important consequences on the type of graph algorithms that can be used since some of these algorithms only work (or are implemented) for a certain type of graph.

In the Netflix graph that we created in the preceding chapter (*Chapter 1, Introducing and Installing Neo4j*), the schema is as follows (all relationships are directed):

- A person acts in or directs a movie, but a movie does not act in a person
- A person is a citizen of a country

This schema has been reproduced in the following diagram:

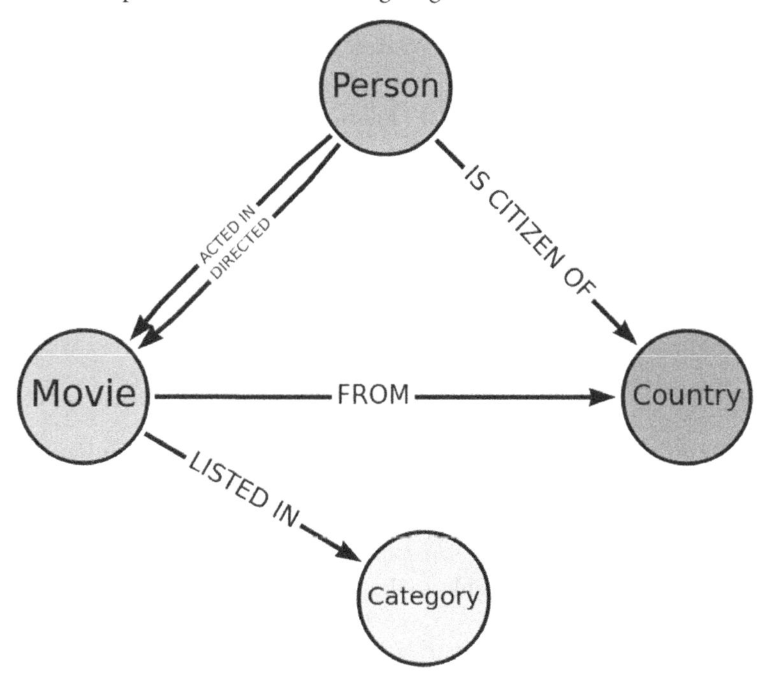

Figure 3.1 – Netflix graph schema

If you have not created the Netflix database yet, a dump with installation instructions has been provided in the code bundle associated with this book, and in the GitHub repository for this book, in the `Chapter03` folder.

Let's add another relationship to our graph: we will materialize the fact that two people know each other if they have worked on the same movie at least once. This is represented in the following graph: both Zion Clark and Floyd Russ participated in the movie titled *Zion*, the former as an actor and the latter as a director. Hence, we can say that Zion Clark and Floyd Russ know each other:

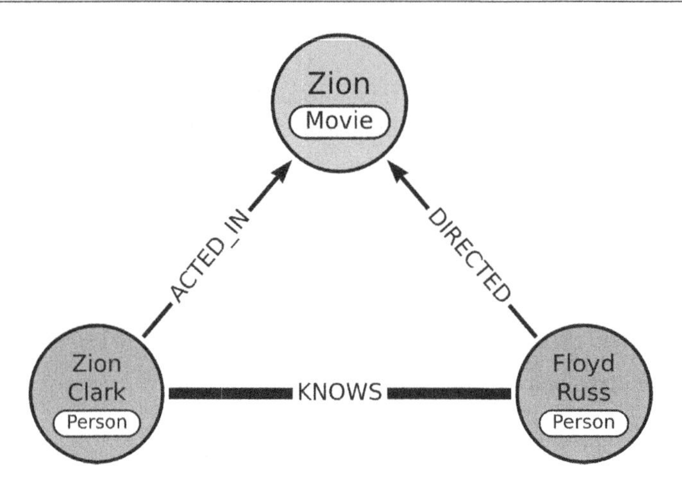

Figure 3.2 – Graph projection

We can create the KNOWS relationships with the following Cypher query:

```
MATCH (p1:Person)
   -[:ACTED_IN|DIRECTED]->(common:Movie)
   <-[:ACTED_IN|DIRECTED]-(p2:Person)
WHERE id(p1) < id(p2)
CREATE (p1)-[:KNOWS]->(p2)
```

The ACTED_IN|DIRECTED pattern means that the relationship must have any of the possible types.

This KNOWS relationship is **undirected**. The id(p1) < id(p2) condition lets us create this relationship only once for the (p1, p2) pair. Without it, we would have to add two relationship objects to the graph: one for the (p1, p2) pair and another one for the (p2, p1) pair, which doesn't make sense here since it carries the same information. Cypher lets us query the graph without specifying the relationship's direction thanks to the -[]- pattern, like so:

```
MATCH (p1:Person {name: "George Clooney"})
   -[:KNOWS]-(other:Person)
RETURN other
```

As we can see, all the returned people have worked on the same movie as George Clooney, either as a director or an actor.

This process of connecting nodes of the same type (Person) by using a graph pattern to connect them, namely here a common (Movie) node and closing the triangle, is a process often used in graph analysis. We'll call this action **graph projection**, and dive deeper into it in the next chapter, where we will use the Neo4j Graph Data Science Library (*Chapter 4, Using Graph Algorithms to Characterize a Graph Dataset*).

Now, let's move on to another important property to consider: the (un)weighted nature of the graph.

Link weight

To determine if a graph is weighted or not, we have to check whether the edges (whether they are directed or not) carry a property that quantifies the strength of the relationship between the nodes. For instance, in the co-authorship graph, a link between two authors can contain the number of papers these two people have collaborated on. In the road network example, edges, representing road segments, can be given a length property so that we know how far two intersections are from each other, which, as you can imagine, can be important for all routing applications.

The following diagram shows an example of a directed weighted graph; the edges are oriented and have weight properties:

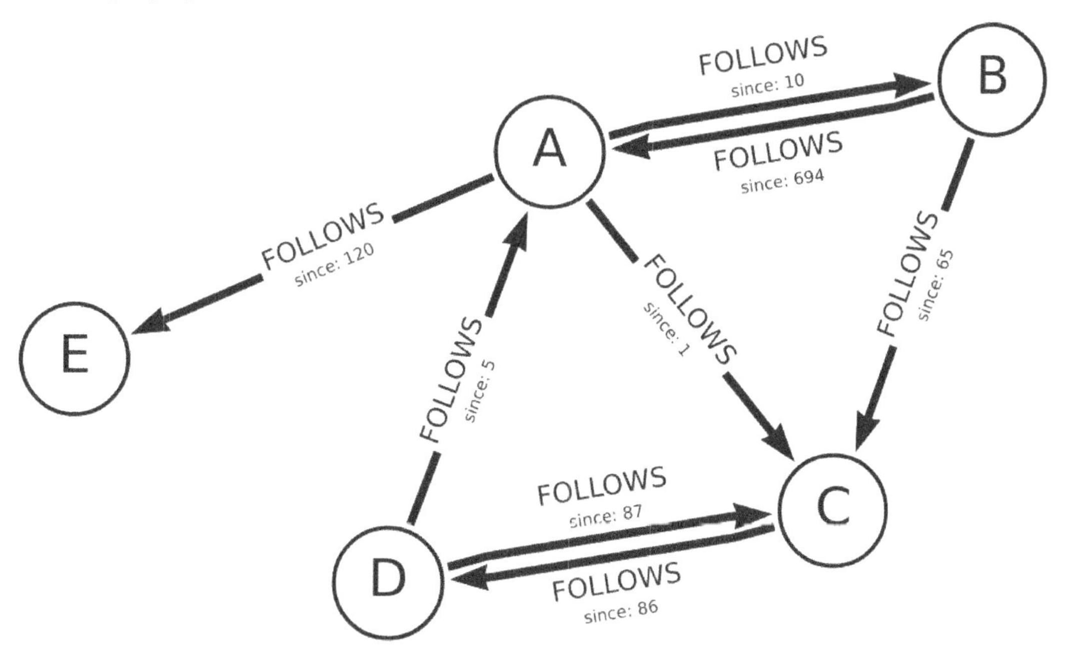

Figure 3.3 – An example of a directed weighted graph

In this graph, nodes represent people who are part of a social network and can follow each other. The `FOLLOWS` relationship is directed, similar to what can be seen on Twitter or Medium, for instance. We also added information to this relationship for storing the number of days this relationship has existed: the higher the `since` property, the longer the relationship between these two people. This property can be seen as a weight, quantifying the strength of the relationship.

Note that a directed graph can be treated as undirected and a weighted graph can always be seen as unweighted if needed, but the contrary is not possible.

Edges are not the only components of a graph to be checked to assess what kind of graph you're facing.

Again, in our Netflix graph, all edges are unweighted so far. But we can add a weight to our `KNOWS` relationship to differentiate between people who worked together once and those who have collaborated on many movies. To do so, let's update our Cypher query:

```
MATCH (p1:Person)
  -[:ACTED_IN|DIRECTED]->(common:Movie)
  <-[:ACTED_IN|DIRECTED]-(p2:Person)
WHERE id(p1) < id(p2)
WITH p1, p2, count(common) as nbMovies
CREATE (p1)-[r:KNOWS_AGG {weight: nbMovies}]->(p2)
```

Our person collaboration network, made up of `Person` nodes and `KNOWS_AGG` relationships, can now be considered weighted. We'll see the consequences of this later in this chapter.

Now, let's move on to the last property we are going to consider here, which is related to the node types the graph contains and the graph's structure.

Node type

Some graphs contain the same types of nodes and are called **homogeneous**. Other types of graphs represent different entities and are **heterogeneous**. In Neo4j, homogeneous graphs have a single node label, while heterogeneous graphs have multiple node labels.

A special kind of heterogeneous graph is a **bipartite graph**. A bipartite graph contains two types of nodes, and links between nodes are only between two nodes of different types. For instance, an e-commerce website with `Product` and `User` nodes and `BUY` links is a bipartite graph (there are no connections between two products or two users, only between a user and a product). An example of a bipartite graph is shown in the following diagram:

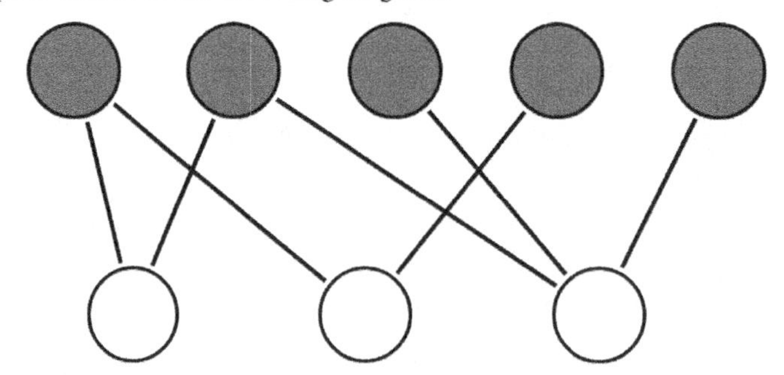

Figure 3.4 – A bipartite graph showing two kinds of nodes. Gray nodes and white nodes are never connected to a node of the same color

Let's consider the Country and Movie nodes in our Netflix graph. These two labels form a bipartite graph: there are relationships from movies to countries, but never from one country to another or from one movie to another. If we consider `Person` and `Movie` nodes, they could also be considered bipartite, until we created the `KNOWS` relationship earlier in this chapter, that connects persons between them, hence breaking the bipartite nature of the graph.

These three elements (edge orientation, weight, and graph composition) are properties that are usually provided with the graph dataset if any. If you are creating a dataset, as we did in the preceding chapter (*Chapter 2, Using Exisiting Data to Build a Knowledge Graph*), it will be up to you to be aware of these characteristics (and document them for future graph analysts using your data – they will be grateful).

But these definitions are not the only ones. Two monopartite, undirected, and unweighted graphs can still differ in many ways. In the following section, we will learn about a node's degree and see what the graph degree distribution tells us about the graph structure.

Computing the graph degree distribution

After the number of nodes and edges, the node's degree is one of the first metrics to compute when studying a new graph. It tells us whether the edges are equally split across nodes or if some nodes monopolize almost all connections, leaving the others disconnected. Now that we've defined the node's degree, we will learn how to compute it with Cypher and draw the distribution using the *NeoDash* graph application.

Definition of a node's degree

The degree of a node is the number of links connected to this node. For undirected graphs, there is only one degree, since we just count all the edges connected to a given node. For directed graphs, we can compute the node's degree in three different ways:

- **Incoming degree**: We count only the edges pointing toward the node
- **Outgoing degree**: We count only the edges pointing outward of the node
- **Total degree**: We count all edges attached to a node, regardless of their direction

Once we have calculated the degree for each node, we can do the following:

- Extract the **minimum degree** of the graph, *G*, denoted $\delta(G)$: If this value is 0, we know that some nodes are disconnected.
- Extract the **maximum degree** of the graph, *G*, $\Delta(G)$: The highest value $\Delta(G)$ can take is $N-1$, where N is the number of nodes. A node with a degree of $N-1$ is connected to every other single node of the graph. If all nodes have a degree of $N-1$, the graph is said to be complete.
- Draw the **degree distribution**: This distribution shows the probability for a node in the graph to have a certain degree. The point, $(k, P(k))$, of the distribution is such that:

$$P_k = \frac{n_k}{N}$$

Here, n_k is the number of nodes in the graph where the degree is k, and N is the total number of nodes in the graph.

Let's go ahead and compute the node degree with Cypher.

In the remainder of this chapter, we will consider the monopartite graph made of `Person` nodes and `KNOWS` relationships only.

Computing the node degree with Cypher

The subgraph we are considering is monopartite (a single node label, `Person`), undirected (`KNOWS` relationships are undirected), weighted graph. It contains 40,948 nodes and almost 358,000 relationships.

To compute the outgoing degree of each node, we are going to use Cypher's aggregation capabilities with the following query:

```
MATCH (n:Person)-[r]->()
RETURN elementId(n) as id, count(r) as degree
```

Let's make sure we understand what this query does:

- In the first line, we start from nodes with the `Person` label and assign them the `n` alias. Then, we follow all outgoing relationships (materialized with the `- ->` pattern), independently of their type and the target node (materialized by the empty `()`), and assign the `r` alias to each of these relationships.
- On the last line, we use the *count* aggregate function to count all relationships, grouping by `id`, which is Neo4j's internal ID for each node. Indeed, in Cypher, the `GROUP BY` clause is implicit; all non-aggregated variables in the `RETURN` clause are assumed to define the groups.

Neo4j's internal IDs

Neo4j's internal IDs are useful for dealing with numbers instead of names, but we should not rely on them for anything other than experimenting and testing since they can be reused when nodes get deleted. Also, note that if you've used Neo4j before version 5, you may have seen the `id()` function, which returns the node's internal ID as an integer. Neo4j 5 introduced the concept of `elementId`, which ensures uniqueness across a cluster, which was not the case for `id()`. Even if we are using a single instance, it is recommended to switch to `elementId` instead.

The first few rows resulting from the preceding query are as follows:

```
╒══════════════════════════════════════════════════════╤══════════╕
│"id"                                                  │"degree"  │
```

```
╞═══════════════════════════════════════════════╪═══════╡
│"4:c8a9e0a0-fea1-4f4c-975b-d47c0d38722d:9046"  │2      │
├───────────────────────────────────────────────┼───────┤
│"4:c8a9e0a0-fea1-4f4c-975b-d47c0d38722d:9047"  │29     │
├───────────────────────────────────────────────┼───────┤
│"4:c8a9e0a0-fea1-4f4c-975b-d47c0d38722d:9048"  │39     │
├───────────────────────────────────────────────┼───────┤
...
```

This means the node with `id=4:c8a9e0a0-fea1-4f4c-975b-d47c0d38722d:9046` has an outgoing degree equal to 2, node `9047` has 29 outgoing relationships, and so on.

Can you guess what should be changed in the preceding query to compute the incoming degree instead of the outgoing degree? Try and think about it for a second before looking at the solution.

Yes, we just need to change the relationship's direction from `-->` to `<--`, like this:

```
MATCH (n:Person)<-[r]-()
RETURN elementId(n) as id, count(r) as degree
```

> **Another option**
>
> If you wanted to write the match pattern as `MATCH ()-[r]->(n:Person)`, that also works – the important thing is that the relationship points `toward` `n`.

Finally, we can also compute the total degree, independently of the relationship's direction, with Cypher. We just have to omit the relationship direction, like this:

```
MATCH (n:Person)-[r]-()
RETURN elementId(n) as id, count(r) as degree
```

Note that there is no < nor > in the relationship pattern, making it undirected. `r` will match both the relationships originating from `n` and the relationships ending at `n`.

These queries give us the degree for every single node. To get to the degree distribution, we have to count the number of occurrences of each degree, which can be achieved with the following Cypher query:

```
MATCH (n:Person)-[r]-()
WITH elementId(n) as id, count(r) as degree
RETURN degree, count(id) as nbOccurrences
ORDER BY degree
```

This query returns a table with two columns: the degree and the number of nodes with such a degree.

You'll agree that this table representation is not super nice and hard to interpret. A visual chart would be easier to understand. This is the reason why we are going to install and use the NeoDash graph application.

Visualizing the degree distribution with NeoDash

NeoDash is a graphical application that can be added to Neo4j Desktop and allows interaction with the active graph (Neo4j Browser is an example of a graph application). NeoDash will fetch the data based on a user-defined Cypher query and display it nicely with many configuration options, from the type of chart to the color scheme and grouping options.

To add a new application to Neo4j Desktop, follow these steps:

1. Click on the **Graph App** icon at the top of the menu bar in Neo4j Desktop.
2. Visit the graph app gallery at `https://install.graphapp.io/` and find **NeoDash**. Click on the last icon install link for the Neo4j sidebar and copy/paste the URL into the **File** or **URL** field in Neo4j Desktop.
3. Click **Install**.
4. Click **Yes** and then **Install** when you're asked to trust the application:

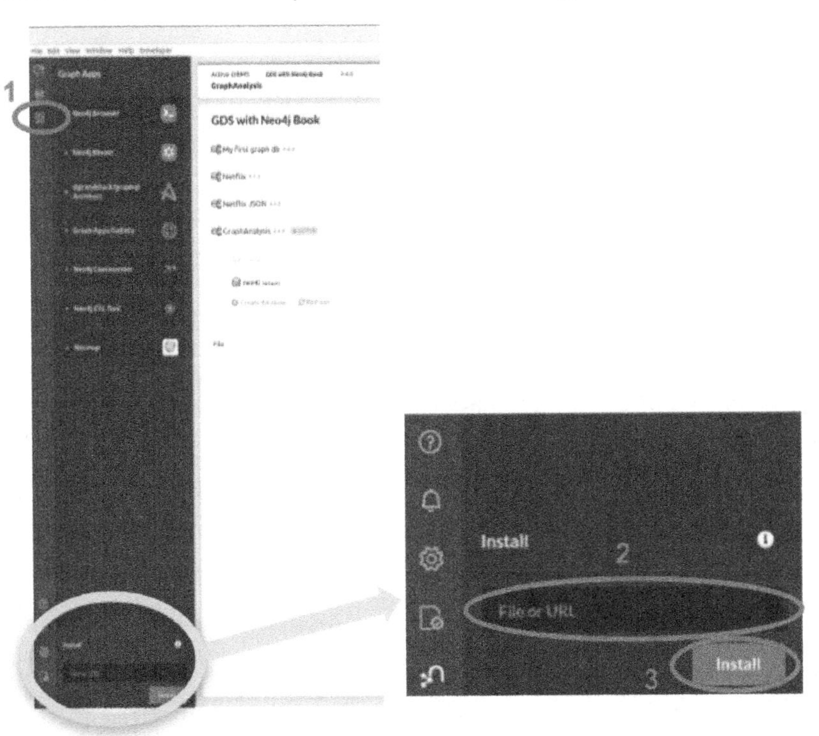

Figure 3.5 – Installing the NeoDash graph application in Neo4j Desktop

Once the app has been installed, it will be listed in the graph app list. You can open it by following these steps:

5. Click on the arrow next to the **Open** button of your running graph.
6. Click on **NeoDash**.

These steps can be seen in the following screenshot:

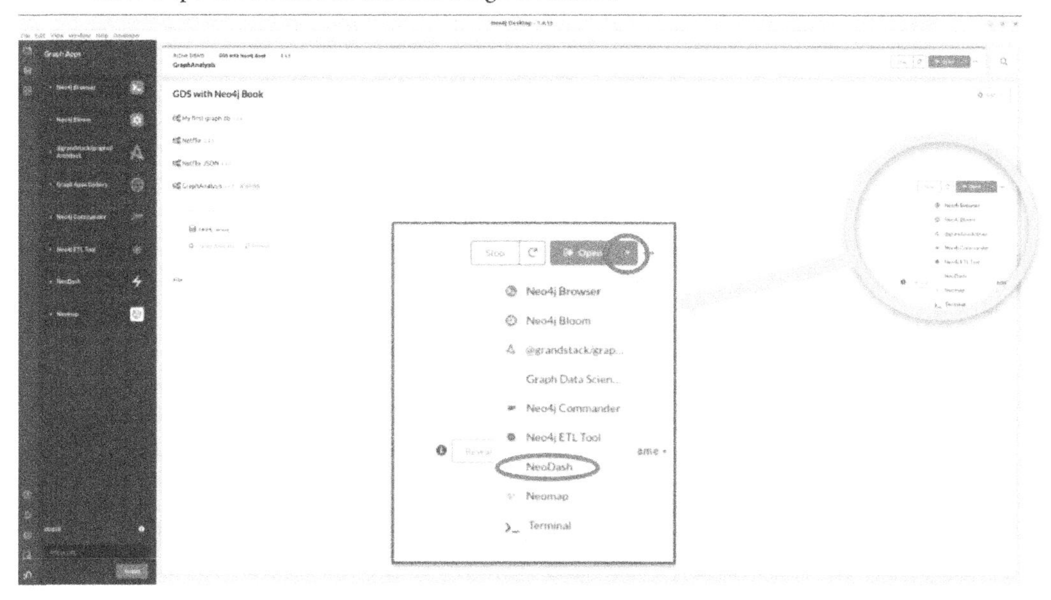

Figure 3.6 – Opening the NeoDash graph application in Neo4j Desktop

The dashboard we are going to create can be seen in the following screenshot:

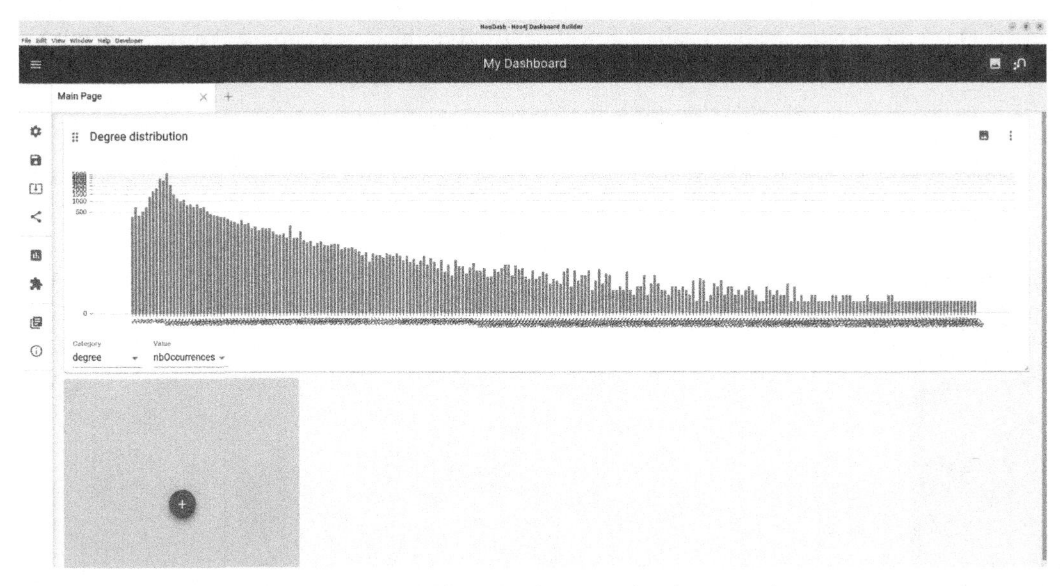

Figure 3.7 – Dashboard to be created with NeoDash

This is quite straightforward, and everything is done via a nice user interface:

7. After starting the application, choose the **NEW DASHBOARD** option.
8. On the displayed dashboard, click on the + button that appears on an empty box.
9. Set the report's name to `Degree distribution`.
10. Click on the three vertical dots on the right-hand side of the name.
11. Set the report's type to **Bar Chart**.
12. Define the Cypher query to fetch data:

```
MATCH (n:Person)-[r:KNOWS]-()
WITH elementId(n) as id, count(r) as degree
RETURN degree, count(id) as nbOccurrences
ORDER BY degree
```

    Note that we are computing the number of occurrences for each degree appearing in the graph so that NeoDash only has to plot these values.

13. Optionally, you can customize the graph by clicking on the **Advanced Settings** switch. For the example you saw earlier, I have set a log scale for the `Y` (value) axis, which you can define by selecting **symlog** for the **Value Scale** property.
14. Click on the disk icon in the top-right corner. That saves your configuration, fetches the data, and renders the chart on the dashboard.
15. Finally, make sure the **Category** and **Value** fields are set to **degree** and **nbOccurrences**, respectively.

Once the chart has been rendered on your dashboard, you can also resize and/or move it at your convenience.

> **Note**
>
> According to the definition of the degree distribution given earlier in this chapter, we are using a non-normalized version here. To obtain the real degree distribution, we'll need to divide the degree by the number of nodes in the graph. This topic will be covered in the *Exercises* section of this chapter.

So far, we have seen the degree distribution of a single graph, which does not tell us much about the diversity of such distributions. To apprehend such diversity, we are going to study another graph. But to do so and to learn something new, we must introduce the Neo4j Python driver.

Installing and using the Neo4j Python driver

We can use the Neo4j Python driver to fetch data from Neo4j and analyze it from Python. In this section, we are going to plot the degree distribution using Python visualization packages.

Counting node labels and relationship types in Python

Let's open the `Neo4j_Driver` notebook (`https://github.com/PacktPublishing/Graph-Data-Science-with-Neo4J/blob/main/Chapter03/notebooks/Neo4j_Driver.ipynb`). To install the Neo4j driver, run the following code in the first cell:

```
!pip install neo4j
```

Let's instantiate the driver and fetch our first bit of data from Neo4j:

1. First, import the required objects:

    ```
    from collections import defaultdict
    from neo4j import GraphDatabase
    ```

2. Then, instantiate a `driver` object, providing it with the connection parameters:

    ```
    driver = GraphDatabase.driver(
        "bolt://localhost:7687",
        auth=("neo4j", "<PASSWORD>")
    )
    ```

3. With the newly created driver, we can start a session and run a query:

```
with driver.session() as s:
    result = s.run("MATCH (n) RETURN n")
```

Here, we are not doing anything with the result variable. But let's parse it to extract the number of occurrences of each label in our graph. This can be achieved with the following code.

4. First, we must define a data container that will hold the result:

```
data = defaultdict(int)
```

5. Then, we can start iterating over the result:

```
for r in result:
```

6. Then, we can access the n variable (the *"n"* is the same as the alias defined in the Cypher query). This variable in Python is of the `Node` type and contains, among others, a property called `labels`. It returns a set of all labels of the node. We iterate over this set to count the number of times we see each label:

```
for l in r["n"].labels:
    data[l] += 1
```

7. Finally, we can print the data variable:

```
print(data)
```

8. Since our graph contains a single node label so far, the preceding print shows the following:

```
{'Movie': 8807, 'Country': 215, 'Category': 42, 'Person':
40948}
```

9. Similarly, to extract the number of relationships per type, we can use the `type()` function:

```
with driver.session() as s:
    result = s.run("MATCH ()-[rel]->() RETURN rel")
    data = defaultdict(int)
    for r in result:
        data[r["rel"].type] += 1
data
```

A difference worth noticing

In Neo4j, nodes can have *none, one, or several labels*, hence the fact that the *labels* property is a set, which can be empty for nodes with no labels. Relationships, on the other hand, must have *one and only one type* so that the `type` property for relationships is not a collection but a simple string, which can never be null.

The result of the previous query is as follows:

```
{
    'FROM': 10019,
    'LISTED_IN': 19323,
    'DIRECTED': 6977,
    'IS_CITIZEN_OF': 349,
    'ACTED_IN': 64124,
    'KNOWS': 357706,
    'KNOWS_AGG': 335862
}
```

The Neo4j Python driver lets you run whatever Cypher query to extract data from Neo4j and process it in Python code. The code snippets we have studied in this section can be reused to understand graph content. However, note that it is not recommended to fetch *ALL* nodes or relationships information for large graphs, for obvious memory reasons.

To go back to our initial topic, we are now going to draw the degree distribution of some graphs stored in Neo4j with Python.

Building the degree distribution of a graph

With our two graphs, let's extract their degree distribution from Python. We will still rely on Neo4j to compute each node degree – after all, this is a graph database, and it's a kind of graph action it is expected to be very good at!

Omitted code

We are not repeating the imports and driver instantiation, similar to the code we have already used in this chapter. However, the full code is available in the `Degree_Distribution` Jupyter notebook (`https://github.com/PacktPublishing/Graph-Data-Science-with-Neo4J/blob/main/Chapter03/notebooks/Degree_Distribution.ipynb`).

Let's start by fetching the degree for each node:

```
def get_degrees(tx, node_label: str, rel_type: str) ->
list[dict]:
    query = f"""
        MATCH (n:{node_label})-[r:{rel_type}]-()
        RETURN elementId(n) as id, count(r) as degree
    """
    result = tx.run(query)
    return result.data()
```

The `data` method returns a list of dictionaries. Let's call this function and see what structure is returned:

```
with driver.session() as s:
    degree_data = s.read_transaction(get_degrees, "Person",
"KNOWS")

print(degree_data[:3])
```

The printed variable is as follows:

```
[{'id': '4:c8a9e0a0-fea1-4f4c-975b-d47c0d38722d:9047',
'degree': 24},
{'id': '4:c8a9e0a0-fea1-4f4c-975b-d47c0d38722d:9048', 'degree':
33},
{'id': '4:c8a9e0a0-fea1-4f4c-975b-d47c0d38722d:9049', 'degree':
11}]
```

We can transform this list of dictionaries into a DataFrame using the standard `DataFrame` constructor:

```
data = pd.DataFrame(degree_data)
```

The `data` DataFrame contains two columns: the node's **id** and its **degree**, as illustrated in the following screenshot:

	id	degree
0	4:c8a9e0a0-fea1-4f4c-975b-d47c0d38722d:9047	24
1	4:c8a9e0a0-fea1-4f4c-975b-d47c0d38722d:9048	33
2	4:c8a9e0a0-fea1-4f4c-975b-d47c0d38722d:9049	11
3	4:c8a9e0a0-fea1-4f4c-975b-d47c0d38722d:9050	11
4	4:c8a9e0a0-fea1-4f4c-975b-d47c0d38722d:9051	8
...	...	...

Figure 3.8 – Content of the data DataFrame

To compute the degree distribution, we just have to count the number of occurrences of each value in the **degree** column. That's exactly what the `value_counts` method is doing:

```
distribution = data.degree.value_counts().sort_index()
```

The `value_counts` method returns a `Series` whose index is the variable (in our case, the degree) and the values are the number of occurrences. We are sorting this `Series` by ascending degree so that the plot looks nicer.

To plot the distribution, we can use the methods included in `pandas` to draw a `Series`:

```
fig = plt.figure(figsize=(12, 8))
ax1, ax2 = fig.subplots(1, 2)
distribution.plot(
        ax=ax2,
        marker="o", color="k",
        title="Degree Distribution",
        xlabel="Degree",
        ylabel="nbOccurrences",
        logy=True,
        logx=True,
    )
fig.tight_layout()
```

Alright – let's look at the results for our first graph, including nodes with the `Person` label. This plot can be seen in the following diagram:

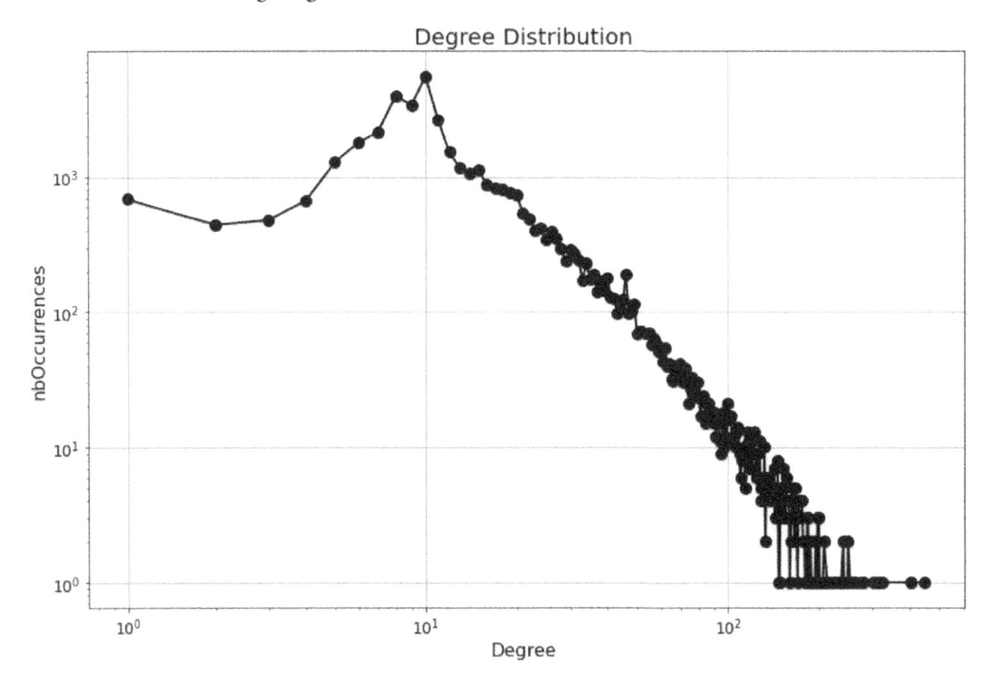

Figure 3.9 – Degree distribution plots for the Person/KNOWS graph

> **Be careful**
>
> This plot was drawn using logarithmic scales for both the `X` and `Y` axes.

What information can we extract from this plot? First of all, you'll notice there seems to be a *maximum*: a lot of nodes have degrees close to 10. As the degree gets closer to 1 or, on the contrary, increases, we have fewer and fewer people. So, the people in our graph have about 10 connections.

The following plot is showing the degree distribution of a different subgraph. Instead of using the `Person` nodes, we use the `Movie` nodes, connecting two movies if at least one person worked in both of them, creating the `RELATED_TO` relationship between two movies (see the *Exercises* section to build the relationship by yourself):

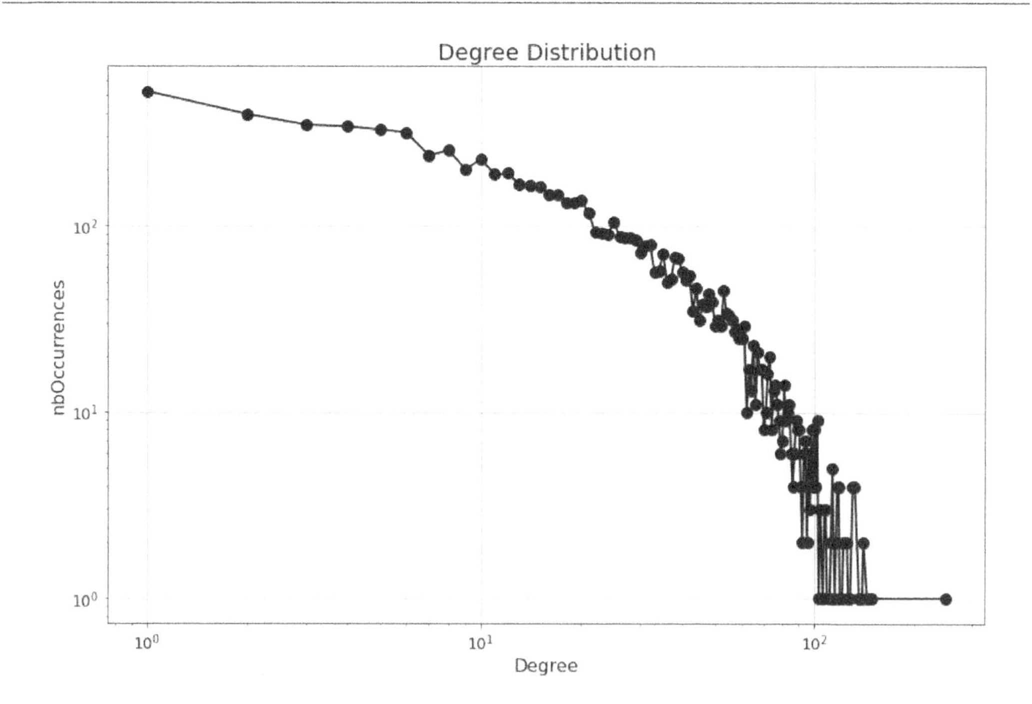

Figure 3.10 – Degree distribution for the Movie/RELATED_TO graph

This time, we can observe that the lower the degree, the higher the number of nodes with such a degree. In other terms, low degrees are more probable than high degrees.

This kind of information can be used if we try and train a link prediction model: can we assume that nodes with a high number of links are more likely to increase their number of connections in the future? Or the other way around? Or maybe there is a *preferred* number of connections, such as for the `Person/KNOWS` graph? We will talk about this problem in *Chapter 9, Predicting Future Edges*.

Before closing this section on degrees, let's focus on two points we've left out so far.

Improved degree distribution

First of all, we are going to think about the impact weight has on degree distribution.

Weighted degree distribution

In the previous section, we drew the **unweighted** degree distribution, meaning the relationship weights are never taken into account – we are just counting the number of relationships. For weighted graphs, such as the one made up of `Person/KNOWS_AGG` entities, we can compute the weighted degree, which is defined by the sum of all incoming/outgoing/total relationship weights. This can sometimes lead to different degree distributions and scales. An exercise at the end of this chapter is dedicated to this calculation.

The special case of nodes with degree 0

The query we have used so far to fetch the node's degree has one major flaw: it won't return nodes with 0 connections. This is because we use a `MATCH` statement with the `(n)-[r]-()` pattern and this pattern doesn't exist for nodes with no relationships, hence these nodes cannot be matched.

We can check this assertion by finding a `Person` node with no `KNOWS` connections. In the following query, we are matching all nodes where the `(:Person)-[:KNOWS]-()` pattern can never be matched:

```
MATCH (p:Person)
WHERE NOT EXISTS((p)-[:KNOWS]-())
RETURN count(p)
```

This tells us that **413** nodes have this characteristic. Let's focus on one of them: **Nurfathiah Diaz**. The following graph shows this person's node (on the right), all connections attached to it (only one `ACTED_IN` relationship), and all connections attached to the only movie this person participated in. As you can see, we do not have more information about this movie – no director and no other actor. That's the reason why this person is not connected through any `KNOWS` relationship:

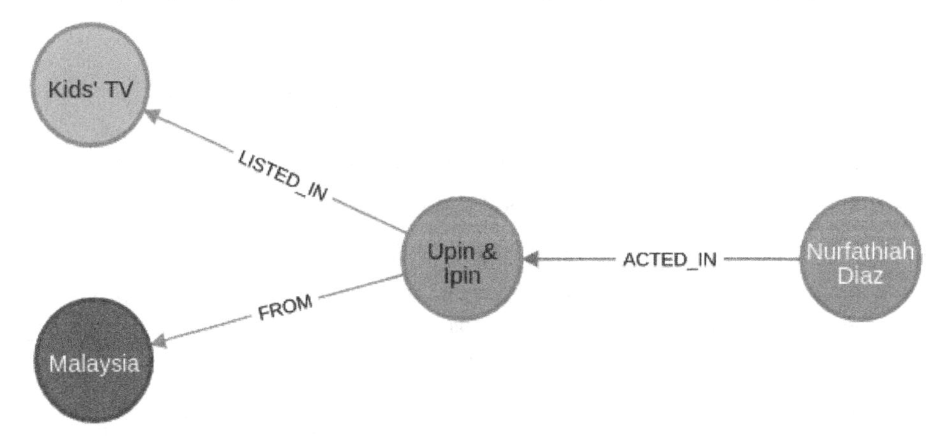

Figure 3.11 – Neighborhood graph for Nurfathiah Diaz

Now, run the following query:

```
MATCH (n:Person)-[r:KNOWS]-()
RETURN elementId(n) as id, count(r) as degree
ORDER BY degree
LIMIT 10
```

You will see that the lowest degree is 1. Indeed, our 413 nodes with no connections are not here!

To fix the query, we can use an `OPTIONAL MATCH` pattern, like this:

```
MATCH (n:Person)
OPTIONAL MATCH (n)-[r:KNOWS]-()
RETURN elementId(n) as id, count(r) as degree
ORDER BY degree
LIMIT 500
```

Here, we `MATCH` all nodes with the `Person` label. The `()-[r]-()` pattern is only an `OPTIONAL MATCH`, meaning it will not *filter* the result further. If the pattern does not exist, then the `r` variable will be null; this is the only difference. This time, we can see our 0-degree nodes in the result:

"id"	"degree"
"4:c8a9e0a0-fea1-4f4c-975b-d47c0d38722d:41559"	0
"4:c8a9e0a0-fea1-4f4c-975b-d47c0d38722d:9823"	0
...	...
"4:c8a9e0a0-fea1-4f4c-975b-d47c0d38722d:10241"	1
"4:c8a9e0a0-fea1-4f4c-975b-d47c0d38722d:9258"	1

Feel free to update the notebook we have used in this chapter to fix the query and add the missing nodes.

As you have seen in this chapter, the degree distribution of a graph can be very different from one graph to another and gives actionable insights about our graph topology and evolution. But the degree is not the only variable that can be computed. We will talk about a few other metrics in the following section.

Learning about other characterizing metrics

The degree is not the only metric that can be computed to characterize a graph. Let's look at a graph detail page on the Network Repository Project (for instance, `https://networkrepository.com/socfb-UVA16.php`). It contains data about the number of nodes, edges, degrees, and other metrics, such as the number of triangles and clustering coefficient.

In the rest of this section, we will provide definitions for some of the metrics listed in the preceding *Figure 3.11*. We will refer to this section in the next few chapters when we use graph-based metrics to build a machine learning model.

Triangle count

The name is self-explanatory, but a triangle is defined by three connected nodes. In a directed graph, edge orientation needs to be taken into account.

For a given node, `n`, its triangle count is found by checking whether its neighbors are also connected to another neighbor of `n`. Look at the following undirected graph:

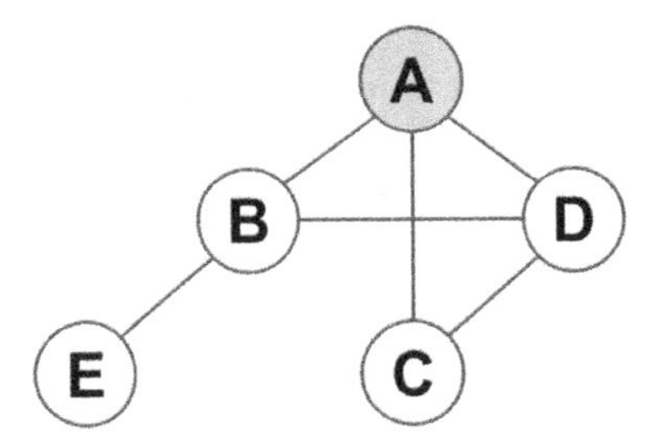

Figure 3.12 – Example graph to compute the triangle count and clustering coefficient

We can compute the number of triangles for node A by starting from the observation that it has three neighbors – B, C, and D:

- B is connected to D, which is also a neighbor of A, forming the triangle **ABD**
- B is connected to E, which is not a neighbor of A; **ABE** is not a triangle
- C is connected to D, which is also a neighbor of A, forming the triangle **ACD**
- Looking at D:
    - It is connected to C, so we find the same **ACD** triangle
    - It is also connected to B, forming the **ABD** triangle we've already counted

We can conclude that node A has two distinct triangles: **ABD** and **ACD**.

By repeating this operation for the other nodes, we will find the following triangle counts:

"node"	"triangles"
"A"	2

```
│"B"   │1          │
├──────┼───────────┤
│"C"   │1          │
├──────┼───────────┤
│"D"   │2          │
├──────┼───────────┤
│"E"   │0          │
└──────┴───────────┘
```

Why are we looking at triangles? It is a very common practice to use triangle observation to make recommendations, for instance. Let's imagine the nodes in the preceding graph represent users of a social network, and the edges represent friendship. From this, we can say that user A is friends with users B and C. There is a certain likelihood for B and C to know each other. If this is the case, we say that we *close the triangle*, formed by the three vertices A, B, and C. On the other hand, D and E, who do not have any common friends, are less likely to know each other in real life.

There are a few metrics that rely on triangle count. The one most frequently used is the clustering coefficient.

Clustering coefficient

We can distinguish between two clustering coefficients: the local one, which is computed for each node, and the global clustering coefficient, which is computed for the whole graph.

Local clustering coefficient

The local clustering coefficient is computed for each node and is defined as follows:

$$C_n = \frac{2T_n}{d_n(d_n - 1)}$$

This is done for a node, n, with a degree, d_n, and triangle count, T_n. Note that each neighbor of n has, at most, d_n-1 other connections, $d_n(d_n-1)$ corresponds to the maximum number of triangles for a node with a degree, d_n, counting each triangle twice. Hence, C_n quantifies how many triangles exist for n, compared to the maximum number of possible triangles given its degree. In other words, we can say that it's a measure of the probability for the neighbors of n to also be connected.

Let's compute the local clustering coefficient for some nodes of the graph, as depicted in the preceding diagram.

Node A has degree 3 and 2 triangles, hence:

$$C_A = \frac{2 \times 2}{3 \times (3 - 1)} = \frac{2}{3} \approx 0{,}67$$

Node C has 2 neighbors and 1 triangle, meaning the following:

$$C_C = \frac{2 \times 1}{2 \times (2-1)} = 1$$

The results for all the nodes have been reproduced in the following table:

"node"	"coeff"
"A"	"0.67"
"B"	"0.33"
"C"	"1.00"
"D"	"0.67"
"E"	"0.00"

Make sure you understand them by following the same approach we used for nodes A and C.

Global clustering coefficient

Similarly, the global clustering coefficient is the proportion of closed triangles over the total number of possible triangles in the network. It provides some information about the overall connectivity of the graph.

Average clustering coefficient

The average clustering coefficient is just the average of all local clustering coefficients over all nodes:

$$LC = \frac{1}{n}\sum_{n} C_n$$

In the next chapter, we will learn how to compute these values with Neo4j by taking advantage of the Graph Data Science Library.

Summary

This chapter taught you some aspects of graph statistics. You now know a few metrics you need to compute when you first start analyzing a new graph, from the number of nodes/edges and the node and edge types to degree-related metrics and distribution.

You also installed the Neo4j Python driver and learned how to extract data from Neo4j to Python and create a DataFrame from data exported from Neo4j.

In the next chapter, we will dig deeper into graph analytics by using unsupervised graph algorithms to learn even more about graph topology. We will learn how to find clusters or communities of nodes in the graph. On the way, we will install and learn about the basic principles of the Neo4j Graph Data Science Library, the plugin we will use intensively in the rest of this book.

Further reading

To better understand some of the concepts that were just approached in this chapter, you can refer to the following resources:

- The best examples for using the Neo4j Python driver can be found in the project's `README` on GitHub: `https://github.com/neo4j/neo4j-python-driver`.
- Vamshi Jandhyala's blog contains valuable inputs about graphs. Regarding this chapter, you can check this blog post, which contains an implementation of the triangle count algorithm in Python, using the `networkx` package: `https://vamshij.com/blog/2020-07-05-triangles-in-graph/`.

Exercises

Challenge yourself with the following exercises related to the content covered in this chapter:

1. Can you imagine an example of a tri-partite graph?
2. Create the `RELATED_TO` relationship between movies that share at least one person (as actor or director).

   Update the Cypher query we used to compute the degree distribution to obtain the normalized degree (divide by the total number of nodes in the graph).
3. Can you draw the weighted degree distribution (total)?

   Hint: The weighted total degree is the sum of all weights of relationships attached to a given node.

4. Advanced: Can you write a Cypher query to compute the triangle count for each node?

   Here is the code to create the small graph we used as an example in Neo4j:

```
CREATE (A:Label {id: "A"})
CREATE (B:Label {id: "B"})
CREATE (C:Label {id: "C"})
CREATE (D:Label {id: "D"})
CREATE (E:Label {id: "E"})

CREATE (A)-[:REL]->(B)
CREATE (A)-[:REL]->(C)
CREATE (A)-[:REL]->(D)
CREATE (B)-[:REL]->(D)
CREATE (B)-[:REL]->(E)
CREATE (C)-[:REL]->(D)
```

4

Using Graph Algorithms to Characterize a Graph Dataset

So far, you have been able to distinguish between different types and topologies of graphs using simple observations and metrics, such as degree distribution. But we can extract more information from a graph structure. In this chapter, we will learn how to find clusters of nodes—or communities—in a network, only based on the nodes and edges in a graph. We will also learn about node importance algorithms, such as PageRank. To do so, we will install and learn the principles of the Neo4j **Graph Data Science** (**GDS**) library, which allows us to run both unsupervised and supervised graph algorithms.

This chapter is a key chapter since lots of the concepts explored herein will be used in the rest of the book, so you are encouraged to stay focused until the end.

In this chapter, we're going to cover the following main topics:

- Digging into the Neo4j GDS library
- Projecting a graph for use by GDS
- Computing a node's degree with GDS
- Understanding a graph's structure by looking for communities

Technical requirements

In order to be able to reproduce the examples given in this chapter, you'll need the following tools:

- Neo4j 5.x installed on your computer (see the installation instructions in *Chapter 1, Introducing and Installing Neo4j*)
- GDS plugin (*version >= 2.2*)
- An internet connection to download the plugins and the datasets

- Any code listed in the book will be available in the associated GitHub repository (`https://github.com/PacktPublishing/Graph-Data-Science-with-Neo4j`) in the corresponding chapter folder

Digging into the Neo4j GDS library

The **GDS** library was first released in 2020. It was the successor of the Graph Algorithm plugin, which first appeared in 2019. Since then, a lot of improvements have been performed in terms of performance and standardization, and a lot of new features have been added, both in terms of algorithm parametrization and new kinds of algorithms. In the following subsections, we give an overview of its content and working principles.

GDS content

As the name suggests, the GDS library contains tools to be used in a data science project using data stored in Neo4j. This includes the following:

- Path-related algorithms
- Graph algorithms
- **Machine learning** (**ML**) models and pipelines
- Python client

Let's talk in a bit more detail about each of these aspects, to understand when and where they are useful.

Path-related algorithms

In graph theory, traversing a graph to find specific paths from one node to another (typically the shortest path) is a key application of graphs. Several algorithms have been developed to find the shortest path between nodes. Dijkstra and A*, to name two, are available in GDS. These algorithms are used for routing, whenever we are talking about vehicle routing on a road network or data packet routing on a computer network.

Graph algorithms

Graph algorithms are mostly unsupervised algorithms that allow us to extract some kind of information from the graph structure itself. It is a broad category containing different types of algorithms. In the GDS library, we can find the following:

- **Centrality algorithms**: Centrality is a measure of node importance in a graph. Different metrics can be used in different circumstances, depending on what *importance* means to you and your data and what types of nodes you want to see in the higher ranks. A node's degree is an example of a centrality metric, where nodes with the highest number of connections have the highest centrality (see *Figure 4.1*). More about this in the *Other centrality metrics* section later in this chapter.

- **Community detection algorithm**: Community detection is the task of forming groups of nodes (or clusters) only looking at the graph structure, meaning the connections. Several techniques have been developed over the years to detect such patterns in a graph. We will detail some of these algorithms in a later section of this chapter (*Understanding a graph's structure by looking for communities*).
- **Similarity algorithms**: In some situations, it can be useful to determine how similar two nodes are in a graph. The similarity between two nodes can be computed based on the graph topology (how many common neighbors they have) or node properties (such as geographic coordinates, in which case we can define similarity as the physical distance between both nodes).

Figure 4.1 shows an example of a centrality measure (namely, PageRank) for the persons in the Netflix dataset we studied in the preceding chapters:

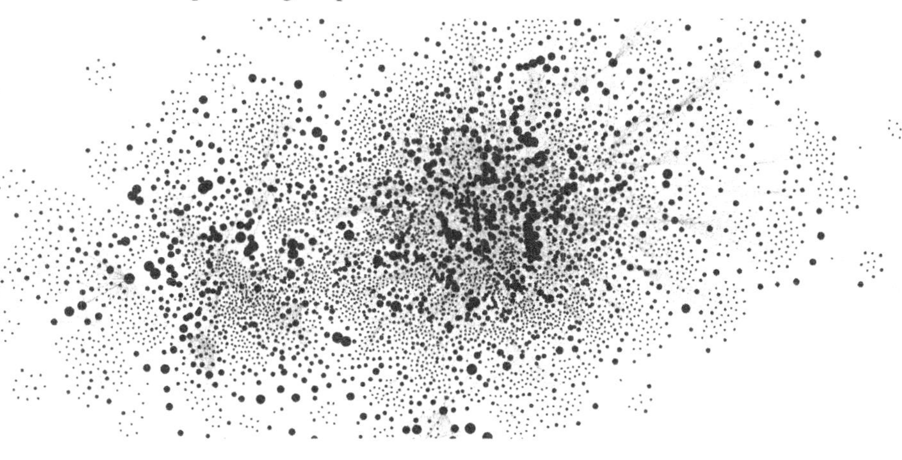

Figure 4.1 – Node importance visualization for the Person nodes in the Netflix dataset

Another aspect of graph analysis and graph ML is the ability to automatically extract features from a graph structure, also known as **graph embedding**.

ML models and pipelines

Embedding is the process of transforming a high-dimensional object, such as text, an image, or a graph, into a low-dimension vector while preserving some key characteristics of the original object, such as meaning for the text or **topology** for the graph. We will discuss this process for a graph in much more detail in *Chapter 7, Automating Feature Extraction with Graph Embeddings for Machine Learning*. Here, I just want to mention that GDS offers this functionality.

Besides implementing very useful embedding algorithms, GDS is also able to do the following:

- Store trained models in a **model catalog** so that the model can be later loaded to make predictions
- Create pipelines and store them in a **pipeline catalog** for node classification or link prediction tasks

These aspects will be covered in *Chapter 8, Building a GDS Pipeline for Node Classification Model Training.*

Last but not least, the Neo4j team has released an official Python client for the GDS library, making it even easier to use graph algorithms for data scientists.

Python client

The GDS library can be used from Cypher, similarly to **Awesome Procedures on Cypher** (**APOC**), by using `CALL` procedures. We are going to learn it this way first in this chapter. But in 2021, a Python client was released, allowing us to call a GDS procedure without writing any Cypher, just Python. We will use this Python client in the next part of this book when building ML models and pipelines since it is a convenient way to integrate GDS with classical ML tools such as Jupyter notebooks and the scikit-learn package.

Before getting started and digging deeper into the library, let's make sure we all know how to install it.

Installing the GDS library with Neo4j Desktop

Installing a new plugin with Neo4j Desktop is quite straightforward. If you followed the previous chapters and have already installed APOC, then you know how to proceed. I remind you of the steps here and in the following screenshot:

1. Find the **Plugins** tab for your graph.
2. Expand **Graph Data Science Library**.
3. Click **Install** and wait. Neo4j Desktop will download the version of the plugin that's compatible with the version of your Neo4j database and restart this database if it was already running so that the new procedures are loaded and ready to use:

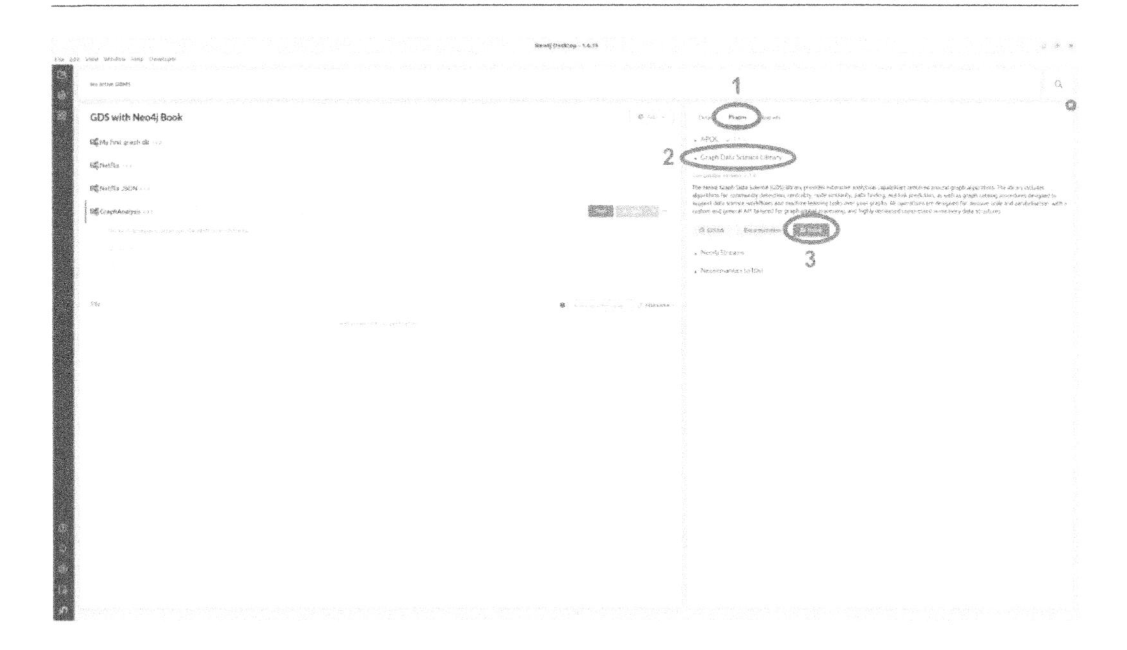

Figure 4.2 – Installing GDS with Neo4j Desktop

GDS/Neo4j compatibility

The GDS version is automatically managed by Neo4j Desktop, but be aware that it won't necessarily install the latest available GDS version due to Neo4j version support from the GDS team, especially if you are using an end-of-life Neo4j version. You can check the compatibility matrix here: `https://neo4j.com/docs/graph-data-science/current/installation/supported-neo4j-versions/`.

You can check the installation worked fine by calling the following procedure from Neo4j Browser:

```
RETURN gds.version()
```

Another interesting procedure is this one:

```
CALL gds.list
```

It returns a list of available procedures from the GDS. As you can see, they all start with the `gds.` prefix so that they are easily identifiable (and, technically speaking, it prevents name collisions between two independent packages). We are not going to discuss each of them (GDS version 2.1.6 contains 338 different procedures!), but the next section is going to explain a typical GDS workflow, and the rest of the book will focus on the most useful applications.

GDS project workflow

When working with GDS, the first thing to do is usually to create a **projected graph**. A projected graph is a graph that will be used to run all algorithms from the GDS. It can be identical to a Neo4j graph, or not. All nodes, relationships, and properties of the projected graph are configurable. You can, for instance, imagine creating a projected graph including the following:

- Only a certain node label(s)
- Only a certain relationship type(s)
- Only certain properties
- New relationships computed on the fly
- New properties computed on the fly

The GDS library provides two ways of creating a projected graph:

- **Native projection**: Nodes, relationships, and properties are selected from the Neo4j database with a standardized configuration
- **Cypher projection**: Entities are filtered from Neo4j or created on the fly, using Cypher queries

We will see in detail how to create a projected graph in the following section.

Another key feature of GDS is to return algorithm results in several ways depending on the needs:

- `stream`: This is the simpler execution mode, where the algorithm results are stored in memory and streamed to the user—for instance, to be displayed in the Neo4j browser or fetched from a Neo4j driver.
- `write`: The algorithm results are not stored in memory but written back to the Neo4j graph as a node property or new relationship, depending on the algorithm.
- `mutate`: In this execution mode, the Neo4j graph is not updated but the projected graph is, which can be useful for chaining algorithms.
- `stats`: Finally, the `stats` mode doesn't return an individual result for each node, but provides overall statistics for the algorithm such as its runtime and the distribution of the computed metric. The exact signature depends on the algorithm.

In addition, there is a fifth mode called `estimate`, which is available for `stream` and `write` modes. It returns estimates of the required memory to run the algorithm. It is useful to check these numbers before any algorithm call when you are working with large graphs in order to make sure you have the necessary resources to perform the computation till the end. These estimates depend on the algorithms since the memory requirements depend on the algorithm and its implementation, so make sure to check the estimations for all of the algorithms you'll need. Also, note that it is possible to estimate the size of the projected graph with the `gds.graph.project.estimate` procedure. In this book, we will be working with relatively small graphs that should not trigger any memory issues; that's why these estimation procedures are only mentioned here but won't be used in the rest of this book.

To summarize this section, look at the following diagram, illustrating a typical workflow of a GDS project:

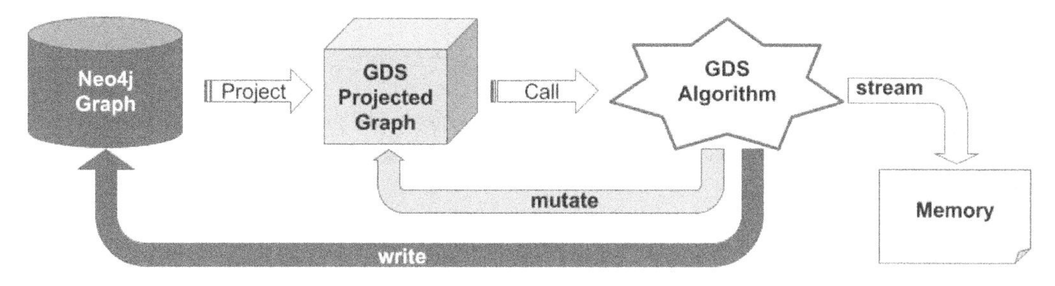

Figure 4.3 – GDS workflow

For all the exercises we are going to work on in this book, we will always do the following:

- Create one or more projected graphs, selecting nodes and relationships from the data stored in Neo4j. *Projected graphs are stored in memory and not persisted.*
- Run one or multiple algorithms on this projected graph, retrieving the results either by streaming them, storing them in the projected graph, or persisting them by writing them back into the Neo4j database.

We have learned in the preceding chapters how to create and feed a Neo4j graph. In the following section, we will learn how to deal with the GDS-related steps, starting from the graph projection.

Projecting a graph for use by GDS

GDS doesn't operate directly on the data stored in Neo4j. Tuned for optimal performance, it uses its own data structure, which can be configured to contain a minimal amount of entities to optimize memory. While your Neo4j graph may contain tens of node labels, each with multiple properties, some algorithms will only use a single node label (for example, `User`) and no property. The GDS library offers the possibility to create a projected graph containing only these nodes. A so-called projected graph can be created using two different procedures:

- `gds.graph.project`: For native projection
- `gds.graph.project.cypher`: For Cypher projection

We are going to detail both of these procedures in the following sections.

Backward compatibility

If you used GDS prior to its 2.0 version, the aforementioned procedures used to be called `gds.graph.create` and `gds.graph.create.cypher`, respectively.

Native projections

In a native projection, we select node labels, relationship types, and properties from the Neo4j graph that will be included in the GDS projected graph. The generic structure of a native graph projection is shown here:

```
CALL gds.graph.project(
      <graphName>,
      <nodeProjectionConfig>,
      <relationshipProjectionConfig>
)
```

Here, the following apply:

- `graphName` is a string containing the projected graph name. This name is important since it will be required by most other procedures so that they know which projected graph they need to use as input.
- `nodeProjectionConfig` and `relationshipProjectionConfig` are strings, lists, or maps containing the required information to filter nodes and relationships.

Let's start by creating our first projected graph and filtering some node labels. We will be working on the graph created during *Chapter 3, Characterizing a Graph Dataset*, containing three node labels: `Person`, `Movie`, and `Country`. The database dump and import instructions are available in the GitHub repository for this book.

Node projection

Let's create our first projected graph, containing only nodes with the `Person` label:

```
CALL gds.graph.project(
      "persons",
      "Person",
      "*"
)
```

The result of the preceding query is reproduced in the following screenshot:

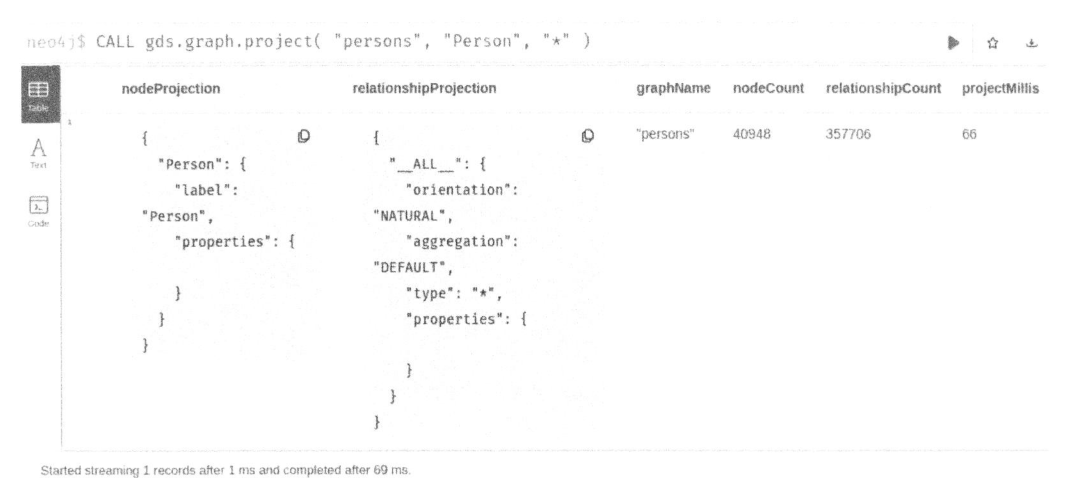

Figure 4.4 – Result of a native graph projection procedure

It contains information about the following:

- **The included nodes and their properties**: Here, only `Person` is included, without any property (notice the empty property map)
- **The included relationships and their properties**: `*` in the procedure call means we are including *all* relationships
- `graphName`, as specified in the procedure `CALL` statement
- **The number of nodes included in the projected graph**: `40948`
- **The number of relationships included in the projected graph**: `357706` (see the next section, *Relationship projection*, for an important point about this number)
- The time required to create this projected graph

In the future, you can find again this information, and more, by calling the following:

```
CALL gds.graph.list
```

This procedure includes graph memory usage and some graph statistics such as degree distribution.

Projected graph lifetime

Beware—a projected graph is *not persisted* and will be lost if you restart your database. So, keep the `CALL` statement to recreate a projected graph in a safe place in order to reproduce your analysis later.

`persons` contains only nodes with the `Person` label. We can include more labels using a list instead of a string in our node projection configuration. For instance, let's create a graph containing nodes with the `Person` *or* `Movie` label:

```
CALL gds.graph.project(
       "personOrMovie",
       ["Person", "Movie"],
       "*"
)
```

As you can see from the formatted node projection and the number of nodes in the projected graph, both `Person` and `Movie` were included:

```
{
  "Movie": {
       "label": "Movie",
       "properties": {
       }
  },
  "Person": {
       "label": "Person",
       "properties": {
       }
  }
}
```

Here, again, the nodes in our projected graph do not have any property, even if they have some in the Neo4j graph. This allows us to reduce the memory usage of the projected graph when these properties are not useful for the analysis we want to perform with it. We will see in a later section how to copy properties from a Neo4j graph to a projected graph, but before that, let's see how to configure relationship projection.

Relationship projection

The relationship projection configuration follows the same pattern as node projection. You can select one relationship type (with a string—for example, `"KNOWS"`), several types (with a list—for example, `["ACTED_IN", "DIRECTED"]`), or be more precise using a map.

There are, however, a few aspects to take into account when selecting relationships.

Orientation

In Neo4j, all relationships are oriented, making a Neo4j graph directed by construction, but some algorithms we are going to use require a projected graph to be undirected. Fortunately, we can control the oriented nature of the relationships we include in a projected graph with the `orientation` property.

The default for projected graphs is to use the same orientation as the one defined in Neo4j—the so-called `NATURAL` projection. It is also possible to use `REVERSE` projection (all relationship orientations are reversed) or `UNDIRECTED` projections (relationships are considered undirected). The different relationship direction configurations are illustrated in the following diagram:

Neo4j Graph

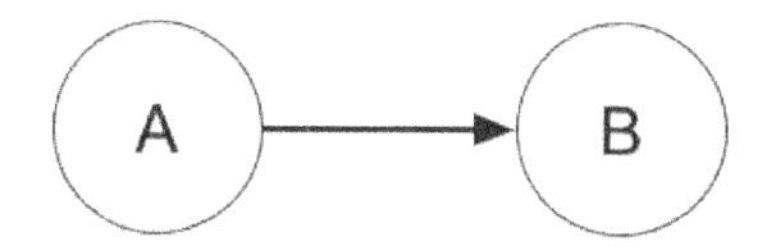

GDS Projected Graph

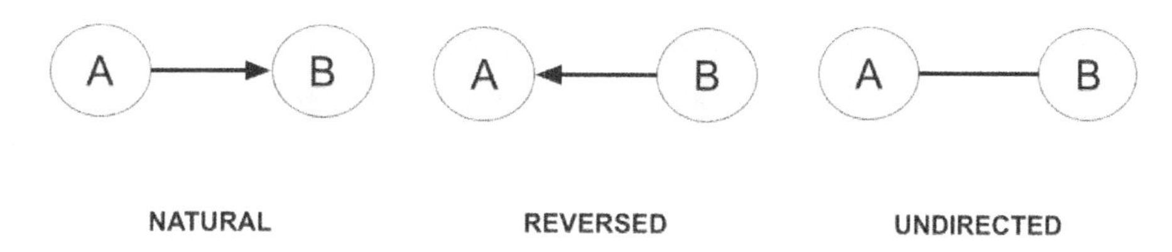

Figure 4.5 – Relationship configuration in a GDS projected graph

Let's study an example and create a projected graph with undirected relationships. We have to specify this requirement in the relationship projection configuration map, like so:

```
CALL gds.graph.project(
      "personsUndirected",
      "Person",
      {
             KNOWS: {
                    orientation: "UNDIRECTED"
             }
      }
)
```

The resulting projected graph properties are reproduced in the following screenshot:

Figure 4.6 – Projected graph with undirected relationships

As you can see, the only change is the number of relationships, going from `357706` (in the default case) to `715412`, exactly double. This means that, internally, GDS is duplicating each relationship to add one in each direction.

> **Note**
>
> In the preceding graph projection, we have not specified the relationship type to be included. By default, GDS selects relationships with a type matching the map key (`KNOWS`, in our case), which will also be the type of the relationships in the projected graph.

Apart from their orientation, there is another important question to ask ourselves about relationships when preparing a projected graph: can we have *parallel* relationships (relationships with the same start and end nodes), and what should we do with them?

Aggregation

The following diagram is an illustration of real data stored in our Netflix graph. As you can see, Steven Spielberg and Harrison Ford have collaborated on more than one movie, leading to the creation of four `KNOWS` relationships between them:

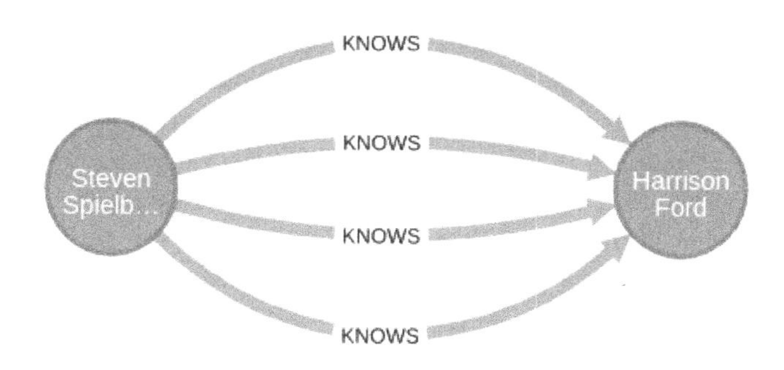

Figure 4.7 – Example of parallel relationships

When creating a projected graph, the default behavior is to keep both relationships, without any property. But you'll realize this is usually not necessary, nor helpful. The easiest way to deal with parallel relationships is to tell GDS to *merge* the relationships and keep only one. This is achieved with the following configuration:

```
CALL gds.graph.project(
      "personsNoParallelRels",
      "Person",
      {
            KNOWS: {
                        aggregation: "SINGLE"
            }
      }
)
```

The `orientation` and `aggregation` options can, of course, be combined in order to create an undirected graph with aggregated relationships.

Included relationships

A native projection only includes relationships when both the start and end nodes are also included by the node projection configuration. You can see this in action with our first projected graph, `projectedPersons`: while we asked it to take all relationships with the `*` symbol, it contains *only* 357,706 relationships, which is the number of relationships between two `Person` nodes. Check it with the following two queries: `MATCH ()-[r]->() RETURN count(r)` to count the total number of relationships in the graph, and `MATCH (:Person)-[r]->(:Person) RETURN count(r)` to count the number of relationships between two nodes with the `Person` label.

There are other ways to aggregate relationships, but we will talk about them in the next section since they will add a new relationship property in the projected graph.

Indeed, so far, we have seen how to select nodes by labels and relationships by types with a particular focus on relationship orientation and aggregation. But the nodes and relationships in our projected graph do not have any attached properties yet. This is not always useful, but some algorithms may need it. Typically, the `weight` property of a relationship can be useful in some circumstances. Considering the graph depicted in *Figure 4.7*, it might be useful to remember how many times two actors have worked together. Many algorithms are able to take such relationship weight into account, which is why we're now going to learn how to add a property to a projected graph.

Node and relationship properties

The syntax to copy a property from Neo4j to a projected graph is similar for nodes and relationships. We have to add a `properties` key in the projection configuration map with the following information:

- Name of the property in the projected graph
- Name of the property in the Neo4j graph
- Optionally, a default value, for nodes that do not have such a property

We will see examples of adding node and relationship properties such as these when appropriate. Indeed, the algorithms we are going to use in the rest of this chapter do not need such properties. However, some algorithms, especially embedding algorithms, are able to take into account both the graph topology (node neighborhood) and node properties to extract a low-dimension vector representation of each node, in which case including node properties in the projected graph becomes important. We'll cover this topic of graph embedding in a later chapter (*Chapter 7, Automatically Extracting Features with Graph Embeddings for Machine Learning*).

That being said, as mentioned earlier, it can sometimes be useful to take into account some kind of weight for relationships. Not all relationships have the same importance. We talked about the relationship between a user and a product, which can be weighted by the number of times the user actually bought the product. But we can also think about street length in a road network, required for any routing system to find the shortest path. We also mentioned earlier the GDS capability to compute the similarity between two nodes, which we may want to store in a relationship between these nodes. In all these cases, the relationship weight is an important property we want to use in our algorithm, and hence we need to add it to the projected graph.

Relationship properties can also be computed by aggregating relationship data when a projected graph is built.

Automatically computed properties for relationships

Relationship weights can be used to measure the importance of a relationship between two persons, assuming for instance that the `KNOWS` relationship is stronger if people have collaborated on multiple movies. We can model this by adding to the projected graph a single `KNOWS` relationship between the two same nodes, but aggregating the number of relationships in Neo4j into a `weight` property.

Technically speaking, this is achieved by adding the `properties` attribute to the relationship projection configuration, like so:

```
CALL gds.graph.project(
    "personsNoParallelRelsWithCount",
    "Person",
    {
        KNOWS: {
            orientation: "UNDIRECTED",
            properties: {
                weight: {
                    property: "*",
                    aggregation: "COUNT"
                }
            }
        }
    }
)
```

The `personsNoParallelRelsWithCount` projected graph contains one single `KNOWS` relationship between each pair of `Person` nodes, adding a property to this relationship in the projected graph only: the number of times such a relationship is seen in the original database.

We can check this by inspecting the content of the projected graph thanks to the `relationshipProperty.stream` procedure:

```
CALL gds.graph.relationshipProperty.stream(
    "personsNoParallelRelsWithCount",
    "weight"
) YIELD sourceNodeId, targetNodeId,
        relationshipType, propertyValue
RETURN sourceNodeId, targetNodeId,
       relationshipType, propertyValue
LIMIT 10
```

Here is an example row returned by this query:

sourceNodeId	**targetNodeId**	**sourceNodeId**	**propertyValue**
9327	14354	"KNOWS"	19.0

Table 4.1 – Relationship properties stored in the projected graph

You can check that the projected graph contains one single relationship for these two node IDs, with a property value equal to 19. We can check that this number is correct by inspecting our initial graph and counting the number of relationships between the source and target nodes, like so:

```
WITH   gds.util.asNode(9327) as source,
  gds.util.asNode(14354) as target
MATCH (source)-[r:KNOWS]-(target)
RETURN count(r)
```

This returns the expected number of 19 relationships. Note that the IDs you see may differ since they correspond to GDS internal IDs, created when a projected graph is created.

Last but not least, we can copy an existing property from Neo4j to a projected graph. To illustrate this feature, we will use the KNOWS_AGG relationship we created in the preceding chapter. It is similar to the KNOWS relationship, except the aggregation has been performed directly in the Cypher query, and the KNOWS_AGG relationship already contains a weight property corresponding to the number of times two persons have collaborated on the same movie. We can create another projected graph with the following query:

```
CALL gds.graph.project(
     "personsKnowAgg",
     "Person",
     {
            KNOWS: {
                   type: "KNOWS_AGG",
                   properties: ["weight"]
            }
     }
)
```

Two differences to note compared to the previous query:

- We rename the relationship from KNOWS_AGG in Neo4j to KNOWS in the projected graph
- We add the weight property to the projected graph

Let's inspect the content of this graph, for the two nodes we studied earlier:

```
CALL gds.graph.relationshipProperty.stream(
     "personsKnowAgg",
     "weight"
) YIELD sourceNodeId, targetNodeId, relationshipType,
```

```
propertyValue
WHERE sourceNodeId = 9327 and targetNodeId = 14354
RETURN sourceNodeId, targetNodeId, relationshipType,
propertyValue
```

This query produces the exact same table as for the `personsNoParallelRelsWithCount` projected graph: the relationship type is `KNOWS`, and the `weight` property value is `19`.

This closes the topic of native graph projection for now. As you can see, we can already configure a lot of a projected graph thanks to configuration maps, both for nodes and relationships, including specifying the behavior of the projected graph for parallel relationships.

But sometimes, native projections are not enough. For instance, we may want to create *virtual* relationships on the fly that only exist in the projected graph. That's where Cypher projections become super useful.

Cypher projections

Sometimes, we want even more control over the entities that will be part of a projected graph. That includes creating relationships on the fly (for example, we want to connect two users who bought the same product, even if this direct relationship doesn't exist in Neo4j) or computing more complex properties (for example, using travel time as relationship weight on a road segment, time that needs to be computed based on segment length, and traffic speed properties).

In order to create a projected graph with Cypher projection, we'll use the `gds.graph.project.cypher` procedure, like so:

```
CALL gds.graph.project.cypher(
     "projectedGraphCypher",
     "<Cypher query to select nodes>",
     "<Cypher query to select relationships>"
)
```

The queries to filter nodes and relationships must return some pre-defined fields:

- For nodes, the query must return a field named `id`, either using Neo4j's `id()` function or any node property that uniquely identifies a node
- For relationships, it must return `source` and `target` fields, containing the ID (as defined in the node query) of the start and end nodes
- All other returned fields are stored as node or relationship properties

Let's take as an example our Netflix dataset, but this time instead of projecting `Person` nodes (black lines on the following diagram), we are going to use the same principle for `Movie` nodes and connect two movies if at least one person has worked in both of them (the gray lines):

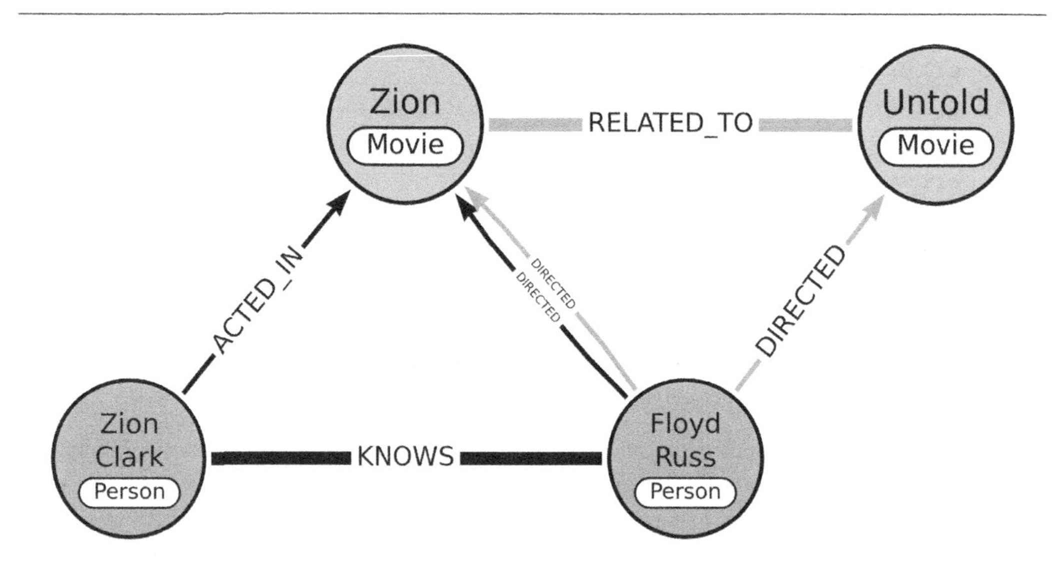

Figure 4.8 – Graph projections

If you have done the exercises in the preceding chapter, the RELATED_TO relationship may already exist in your database, but for the sake of the example, we will pretend it doesn't. The Cypher query to fetch the source and target nodes, together with the force of the relationship, is the following:

```
MATCH (m1:Movie)
  <-[:ACTED_IN|DIRECTED]-(common:Person)
  -[:ACTED_IN|DIRECTED]->(m2:Movie)
WITH m1, m2, count(common) AS nbCommonPersons
RETURN
  id(m1) AS source,
  id(target) AS target,
  nbCommonPersons AS weight
```

The final query to build such a projected graph is shown here:

```
CALL gds.graph.project.cypher(
     "projectedGraphCypher",
     "MATCH (n:Movie) RETURN id(n) as id, n.releaseYear as
year",
     "MATCH (m1:Movie)<-[:ACTED_IN|DIRECTED]-(common:Person)-
[:ACTED_IN|DIRECTED]->(m2:Movie) WITH m1, m2, count(common)
AS nbCommonPersons RETURN id(m1) AS source, id(m2) AS target,
nbCommonPersons AS weight"
)
```

Using the `nodeProperty.stream` or `relationshipProperty.stream` procedures to inspect the projected graph content, you'd find the `year` (for nodes) and `weight` (for relationships) properties created from Cypher.

You should now have a better understanding of graph projection in the GDS context and understand the possibilities offered by such a design decision. Don't worry if you feel a bit overwhelmed with the information in this section. We will see more examples of graph projection in the rest of this book, and the concept will become clearer and clearer to you.

In the following section, we will dig into algorithm run modes, using a node's degree GDS procedure as an example.

Cleaning our graph catalog

In the last few sections, we have created quite a few projected graphs that are now only taking up memory space. We can clean them either by restarting the Neo4j database (since they are not persisted) or by running the following query, which will drop all projected graphs:

```
CALL gds.graph.list() YIELD graphName as gn
CALL gds.graph.drop(gn) YIELD graphName
RETURN graphName
```

In the following sections, we will only use one projected graph, as illustrated here:

```
CALL gds.graph.project(
      "persons",
      "Person",
      "*"
)
```

Let's continue our exploration of GDS by computing a metric we studied in the preceding chapter: a node's degree.

Computing a node's degree with GDS

We have studied the node degree metric and its distribution in the preceding chapter, *Chapter 3, Characterizing a Graph Dataset*. At that time, we computed the node's degree using a Cypher query. GDS provides a procedure to perform the same computation, on a projected graph. We are now going to use this procedure, whose results are well known, in order to understand the different algorithm modes and configuration options.

All algorithm procedures from GDS use the same syntax:

```
gds.<algoName>.<executionMode>(<graphName>,
<algoConfiguration>)
```

Here, the following applies:

- `algoName` is the name of the algorithm. Note that some algorithms are included in an alpha or beta version, in which case they are accessible via `gds.alpha.<algoName>` or `gds.beta.<algoName>`.
- `executionMode` is one of `stream`, `write`, `mutate`, `estimate` or `stats`, as defined in the *GDS project workflow* section.
- `graphName` is the name of the projected graph (that must already exist) on which the algorithm will be run.
- `algoConfiguration` is an algorithm-specific configuration map.

Let's see all of this in action with the degree procedure, starting from the `stream` mode.

stream mode

Calling the degree procedure from GDS in `stream` mode is, then, as simple as this:

```
CALL gds.degree.stream("persons")
```

The result must look like this:

```
╒════════╤═══════╕
│"nodeId"│"score"│
╞════════╪═══════╡
│9046    │0.0    │
├────────┼───────┤
│9047    │24.0   │
├────────┼───────┤
│9048    │33.0   │
├────────┼───────┤
...
```

> **Two types of node IDs**
>
> In this chapter, we are talking about multiple IDs for a single node—pay attention to make a clear distinction between them. The returned `nodeId` property is a *GDS internal ID*, assigned to each node in a projected graph. Being an internal property, `nodeId` *can be different from one machine to another*, meaning you may have results different from mine because your mapping might be different. This value is different from the one returned by the `id()` function of Neo4j, which shows *Neo4j's internal node ID.*

Fortunately, we can transpose the opaque `nodeId` property to a Neo4j node, using the `gds.util.asNode` function:

```
CALL gds.degree.stream("persons")
YIELD nodeId, score
WITH gds.util.asNode(nodeId) as node, nodeId, score
RETURN node.name as name, nodeId, score
LIMIT 100
```

Here, `node` is the node object stored in the Neo4j graph, meaning we can access its properties. In our graph, nodes have a single property called `id`. So, the result of the previous query is this:

```
╒════════╤═══════════════════════════════════════════╤═════════╕
│"nodeId"│"name"                                     │"score"  │
╞════════╪═══════════════════════════════════════════╪═════════╡
│9046    │"Kirsten Johnson"                          │0.0      │
├────────┼───────────────────────────────────────────┼─────────┤
│9047    │"Julien Leclercq"                          │24.0     │
├────────┼───────────────────────────────────────────┼─────────┤
│9048    │"Mike Flanagan"                            │33.0     │
├────────┼───────────────────────────────────────────┼─────────┤
...
```

Put apart the ordering, these results must be identical for your graph since we are now viewing the `name` property that we set when, and this name is deterministic. That means that if you check the computed degree for `"Mike Flanagan"` on your machine, it must be equal to 33.

The YIELD keyword

In Neo4j, the `YIELD` statement after a `CALL` statement can be omitted if `CALL` is the only statement of the query. As soon as you need anything else, such as a `WITH` statement and a `RETURN` statement in the preceding query, the `YIELD` statement becomes mandatory.

To know which variables are yielded by a procedure, you have two options to find out the answer: check the online documentation (for instance, for the degree procedure here: `https://neo4j.com/docs/graph-data-science/current/algorithms/degree-centrality/#algorithms-degree-centrality-examples-stream`) or check the procedure signature in the included documentation by calling `CALL gds.list("degree.stream")`. The preceding query will show you the procedure signature, where you can identify the returned values: `(nodeId :: INTEGER?, score :: FLOAT?)`.

Streaming is useful for relatively small graphs when the result can be held in memory without breaking your computer. It is the mode to use when you want to see the results directly in Neo4j Browser or when you need to fetch the results from the Neo4j Python driver, for instance. We'll see a few examples in the coming chapters. But when the graph is large, streaming results is not possible. That's when the `write` mode becomes essential.

write mode

To use a graph algorithm in `write` mode, we need to do the following:

- Use the `gds.<algo>.write` procedure
- Specify the `writeProperty` value, a.k.a. the name of the property that will store the algorithm results into Neo4j

As an example, let's see how to compute a node's degree and store the results in a property called `"degree"` in Neo4j:

```
CALL gds.degree.write(
  "persons",
  {
    writeProperty: "degree"
  }
)
```

> **Warning**
> The `writeProperty` parameter we added in the configuration map is mandatory; omitting it will result in the following self-explanatory error message: `IllegalArgumentException: No value specified for the mandatory configuration parameter `writeProperty``.

The result of the preceding query is illustrated in the following screenshot:

Figure 4.9 – Result of the degree procedure in write mode

As you can see, it does not contain the degree for each node, but an overview of the degree distribution with percentiles, and some algorithm statistics.

We can check that the `write` mode actually worked and a new property was added to each node with the `Person` label (due to the projected graph only containing these nodes) with the following query:

```
MATCH (n:Person)
RETURN n.name as name , n.degree as degree
LIMIT 10
```

The result of the preceding query demonstrates that the `degree` property has properly been added to the graph (you can remove the `LIMIT` clause if you doubt it, or, even better, count the number of `NULL` values to find out if it is equal to 0).

There is a last important mode for algorithms, especially useful when algorithms need to be chained. In the current scenario, if you want to use the output of an algorithm as input for another one, you'll have to do the following:

1. Create a projected graph.
2. Run your first algorithm and store the results in Neo4j.
3. Create a second projected graph and add a new property.
4. Run your second algorithm.

This process is not optimized, and, fortunately, GDS offers the possibility to update a projected graph via `mutate` mode.

mutate mode

In order to call an algorithm in mutate mode, we just need to use the following code:

```
CALL gds.degree.mutate(
  "persons",
  {
    mutateProperty: "degreeMutate"
  }
)
```

After running this query, the degreeMutate property does not exist in the Neo4j graph. You can check this assertion with the following Cypher query:

```
MATCH (n:Person)
RETURN n.name as name, n.degreeMutate as dm
LIMIT 10
```

You'll notice that the dm column is always NULL, and removing the LIMIT value won't change this observation.

However, checking the content of the projected graph will reveal that this property has actually been saved there. That's the goal of this query:

```
CALL gds.graph.nodeProperty.stream(
  "persons",
  "degreeMutate"
)
```

For each node in the projected graph, it shows the value of the degreeMutate property, which is not null.

> **Good practice**
>
> The nodeProperty.stream procedure to inspect the content of a projected graph is to be used mainly for debugging. As soon as a property needs to be read from outside GDS, it must be retrieved through a stream or write algorithm mode.

We have talked a lot about projected graphs and the three main algorithm run modes (stream, write, and mutate). Before closing our tour of the GDS library, let's talk a little bit about algorithm configuration—it will be an opportunity for us to understand some subtleties of graph orientation.

Algorithm configuration

Do you remember from the preceding chapter the different degree definitions? We have the incoming degree (number of relationships pointing to a given node), the outgoing degree (number of relationships leaving a given node), and the total degree (total number of relationships attached to a node, regardless of their direction).

With the preceding queries (whichever mode is used), which of these degrees do you think we were computing?

To answer this question, we need to focus on two things:

- **The projected graph configuration**: In the `persons` projected graph, we are using the default configuration for relationships, meaning the relationships are oriented using the natural orientation (that is, the one defined in Neo4j).
- **The algorithm configuration**: Some algorithms, such as the one we are using right now, also offer the possibility to tune relationship orientation at runtime. The default value for this parameter is to use the `NATURAL` orientation, which in this case means natural with respect to the projected graph.

So, using the default configurations everywhere, we end up using oriented relationships, and we are counting the number of outgoing relationships, so this is the outgoing degree (equivalent to the `-->` Cypher pattern).

If we want to compute the total degree, independently of relationship orientation, here is one way to proceed, using the same projected graph by telling the algorithm to ignore relationship orientation:

```
CALL gds.degree.stream(
     "persons",
     {
           orientation: "UNDIRECTED"
     }
)
YIELD nodeId, score
WITH gds.util.asNode(nodeId) as node, score
RETURN node.name as name, score
LIMIT 100
```

This is an illustration of the importance of a cautious definition of graph projection and algorithm configuration. Depending on the algorithm, more parameters are of course available. For instance, it is possible to restrict the node labels used in one particular algorithm if the projected graph contains multiple labels. Finally, some algorithms have *tuning parameters*. All of these configurations will be passed to the algorithm through the configuration map, as we did with the `orientation` parameter in the preceding query.

The last sections have given you an overview of GDS, from projected graph to algorithm run mode and configuration, by studying a metric you already knew: the node degree distribution. It is time to discover other kinds of metrics included in the GDS library, first focusing on other ways to measure node importance, also known as **centrality measures**.

Other centrality metrics

Computing a node's degree is a measure of the node's importance in the graph: the higher the number of connections, the more important the node. But it is still a very simple algorithm, and more sophisticated ones have been elaborated. We will discuss two of them in this section: **PageRank** and the **betweenness** centrality.

PageRank

PageRank is named after Larry Page, co-founder of Google, who invented this algorithm to rank search results from the most relevant to the less relevant. His idea was that not all neighbors add the same value: one node is connected to six other nodes, but each of these neighbors having no other connection (node **A** in the following diagram) is maybe less important than a node with two neighbors, each having four connections (node **B**)? If this is true that considering only one hop, node **A** can reach more nodes, this is not true anymore when considering a two-hop path, where node **A** can still reach six nodes, but node **B** reaches ten nodes—its two neighbors plus its eight neighbors:

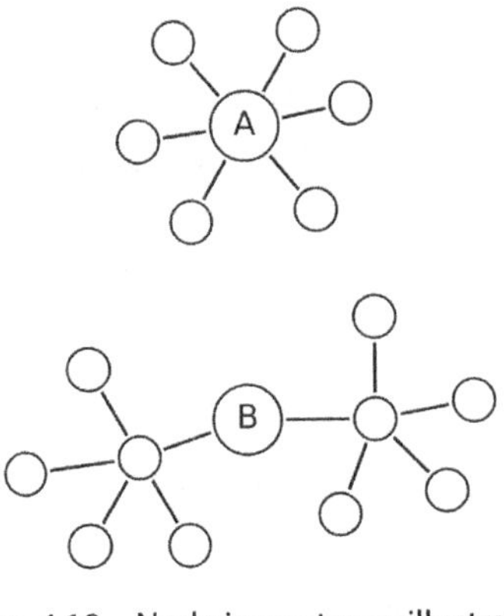

Figure 4.10 – Node importance illustration

The PageRank algorithm is an iterative algorithm that will update each node's importance based on its neighbors' importance until a certain level of stability is reached (or a certain number of iterations). With GDS, it is available as follows:

```
CALL gds.pageRank.write(
        "persons",
    {
```

```
        writeProperty:   "pr"
        // maxIterations: 100
    }
)
```

In practice, PageRank is used more often than the degree centrality, for the reasons mentioned previously.

There are other cases for which neither the degree nor PageRank centrality would convey the information we are looking for. So far, we considered a node as important if it is densely connected, or part of a densely connected region and can help spread information faster. But in a road or computer network, for instance, some nodes are critical not because they have many connections but because if they are removed from the network due to closed roads or breakdown, the entire network communication might thus be affected.

Betweenness

Consider again node **B** in the preceding diagram. What happens if we remove it from the graph? We would end up with the graph depicted in the following diagram, with two totally disconnected parts:

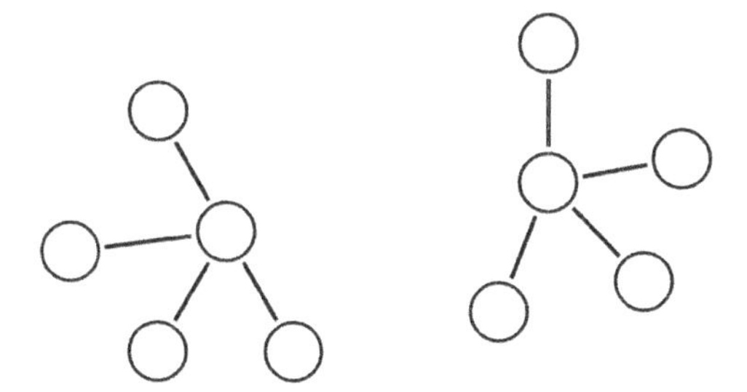

Figure 4.11 – What if we remove node B from the previous graph?

If the communication between those two parts is important, we would be in severe trouble. In such cases, node **B** is probably the node we'd want to protect as much as possible.

In such cases, the *betweenness* centrality is a more suitable metric to measure node importance. It basically counts the number of paths going through a given node: the higher the number, the more important the node:

```
CALL gds.betweenness.write(
        "persons",
    {
        writeProperty:   "btw"
    }
)
```

Degrees and centralities are one way to understand the graph, by spotting important nodes depending on what importance means to you and your data. But we can extract even more information from a graph topology. In the coming section, we are going to investigate the community structure of our graph.

Understanding a graph's structure by looking for communities

In a graph, the repartition of edges is often a key characteristic. Indeed, graph traversal is used by many algorithms to propagate some values from one node to its neighbors, until some equilibrium is reached. Knowing in advance that some groups of nodes are totally isolated from, or share very few links with, the rest of the graph is key information to understand the result of such algorithms. Besides those technical details, the knowledge that some nodes tend to be *more connected* with each other with respect to other nodes in the graph, forming a community can also be used as an input feature for an ML model. You can, for instance, imagine finding communities in your user base depending on the products they frequently buy and identifying the group of coffee aficionados, different from the group of tea lovers, that will get different recommendations.

Number of components

The next goal of this analysis is to try to find whether there are *isolated* groups, meaning groups of nodes not connected to the rest of the graph. The algorithm allowing us to retrieve such information is the **connected component** algorithm.

The algorithm signature, as written in GDS's manual, is the following:

```
CALL gds.wcc.stream(
  graphName: String,
  configuration: Map
)
YIELD
  nodeId: Integer,
  componentId: Integer
```

That means that, for each node, the algorithm returns a number, the `componentId` value, representing the community the node has been found to belong to.

Let's run this algorithm on our graph. We can call the algorithm and stream the results. But in order to better understand the graph structure, we are not going to look at the raw result (the community ID for each node) but aggregate it to count the number of nodes in each community:

```
CALL gds.wcc.stream(
  "persons"
```

```
)
YIELD nodeId, componentId
WITH componentId, count(nodeId) as numberOfNodes
RETURN componentId, numberOfNodes
ORDER BY numberOfNodes DESC
```

The result of the preceding query is reproduced here:

"componentId"	"numberOfNodes"
1	36878
848	34
902	26
...	
2923	1
2928	1

componentID

`componentId` (or `communityId` in other algorithms) values are not necessarily sequential nor carry any particular meaning; they are just the result of the algorithm's implementation. In other words, it doesn't matter that the `componentId` value is `0`, `1`, or `2123`; the only relevant information is whether *node X and node Y* have the same `componentId` value or not.

Here is what we can say from these results. The algorithm identified different components:

- One big component, whose ID is `1`. It contains `36878` nodes, representing 90% of all persons.
- A few other components with a few tens of nodes.
- A lot (413) of components containing a single node.

These 413 nodes correspond to totally isolated nodes, meaning nodes not connected to any other person, as we can check with the following Cypher query, which returns the exact same number of nodes (413):

```
MATCH (p:Person)
WHERE NOT Exists((p)--(:Person))
RETURN count(p)
```

(We could check that the nodes are indeed the same by checking the names.)

The connected components algorithm gives us an indication of the graph structure. Identifying disconnected components like this is crucial since some algorithms rely on information propagation through relationships, such as PageRank. If some nodes are not connected to others, the algorithm may not function as expected.

This algorithm, while useful, is a bit extreme in the sense that it will only find out components sharing *zero* relationships. However, in a graph with no disconnected components, it is still possible to create groups of nodes—or clusters. The GDS library contains several algorithms whose goal is to identify communities. We will detail in the next section the Louvain algorithm, based on a metric called **modularity**.

In order to ease the analysis and make it faster, we are now going to study only the internal structure of the biggest connected component. To do so, we're going to add a specific label to nodes belonging to this component, as follows:

1. Run **Weakly Connected Component** (**WCC**) in `write` mode:

   ```
   CALL gds.wcc.write(
     "persons",
     {writeProperty: "wcc"}
   )
   ```

2. Identify the biggest WCC component and add the `MainComponent` label to nodes in this community:

   ```
   MATCH (p:Person)
   WITH p.wcc as cid, count(p) as nbPerson
   ORDER BY nbPerson DESC LIMIT 1
   MATCH (pp:Person {wcc: cid})
   SET pp:MainComponent
   ```

3. Create a projected graph for nodes in the main component only:

```
CALL gds.graph.project(
      "mainPersons",
      "MainComponent",
      {
       KNOWS: {
             orientation: "UNDIRECTED"
            }
      }
)
```

Let's go ahead and study the internal structure of this big community.

Modularity and the Louvain algorithm

How can we find communities in a connected graph? Let's take the problem from the other end: imagine you are given a graph partition; how would you determine if this partition is a good one?

There is no unique answer to this question. One of the metrics we can think of is the following: if each node shares more relationships with nodes in the same community rather than the expected relationships for a random graph, we can say we have a good partition. This metric has been used a lot in graph analysis and is called **modularity**.

There are different algorithms whose goal is to find an optimal graph partition based on modularity. Here, we will use the **Louvain algorithm**. In GDS, it lies under the `louvain` procedure. Let's use it on a graph made of nodes with the `MainCommunity` label only. Similarly to the previous analysis, print the number of nodes per community:

```
CALL gds.louvain.write(
  "mainPersons",
   {writeProperty: "louvain"}
)
```

In the second step, we can analyze the written property:

```
MATCH (p:MainComponent)
WITH p.louvain as louvain, count(p) as numberOfNodes
RETURN louvain, numberOfNodes
ORDER BY numberOfNodes DESC
```

The first rows of the result are reproduced here:

```
╒═════════╤═══════════════╕
│"louvain"│"numberOfNodes"│
╞═════════╪═══════════════╡
│1879     │4358           │
├─────────┼───────────────┤
│22421    │3802           │
├─────────┼───────────────┤
│32213    │3282           │
├─────────┼───────────────┤
│21074    │1397           │
├─────────┼───────────────┤
...
```

As you can see, despite our nodes having connections with each other, many different communities are identified, containing between 1 and more than 4,000 nodes. Can we understand these communities?

Making sense of discovered communities

It is reasonable to assume that persons coming from the same country are more likely to work together and build strong connections, due to geographical closeness and sometimes no language barrier. Remember from *Chapter 2, Using Existing Data to Build a Knowledge Graph,* in the section *Importing Data into Neo4j to Build a Knowledge Graph,* or from the graph schema shown in *Figure 4.1*, we can fetch the country of citizenship for each person.

If you do not have the Netflix database yet, you can create it by downloading the database dump and following the instructions here: `https://github.com/PacktPublishing/Graph-Data-Science-with-Neo4J/tree/main/Chapter03#import-database` (from *Chapter 3, Characterizing a Graph Dataset*).

We have seen in *Chapter 2, Using Existing Data to Build a Knowledge Graph,* in the section *Importing Data into Neo4j to Build a Knowledge Graph,* how to fetch data from Wikidata to save the country of citizenship for each person, but fetching all data from Wikidata is time-consuming, hence I provide a dataset that's ready to use: `Chapter04/data/person_citizenship.csv`. You can load it with the following query:

```
// first delete existing relationships
MATCH ()-[r:IS_CITIZEN_OF]-()
DELETE r
```

```
// then load new ones
LOAD CSV WITH HEADERS
FROM "file:///person_citizenship.csv" AS row
MATCH (p:Person {name: row.person})
MERGE (c:Country {name: row.country})
CREATE (p)-[:IS_CITIZEN_OF]->(c)
```

Let's now reuse the Neo4j Python driver to fetch the data and analyze it in Python:

```
from neo4j import GraphDatabase
driver = GraphDatabase.driver("bolt://localhost:7687",
auth=("neo4j", "<PASSWORD>"))
```

Since Neo4j 5, we can directly export a Cypher query result into a pandas DataFrame like this:

```
with driver.session() as s:
     result = s.run("""
           MATCH (p:MainComponent)-[:IS_CITIZEN_OF]-
>(c:Country)
           RETURN p.name as personName, p.louvain as
louvainCommunity, c.name as countryName
""")
     df = result.to_df()
```

Here are two other examples of real-life scenarios in which these communities could be useful:

- **In an e-commerce context, where your graph nodes are users**: Two users are connected if they have bought the same product. In this case, communities of users could represent users with common interests (running, fishing, cinema, cooking…) and help in improving product recommendations by extending the products of interest to the ones bought by someone in the community, even if users are not directly connected with each other.
- **In an enterprise graph, where each node is an employee and each relationship represents an email exchange between employees**: Finding communities there might help you identify groups of people who are more used to working together, even in a different department, and form more effective teams for coming projects.

These are just two examples to give you some hints about why communities are important. They not only help you understand the data but they can also be used to make decisions. We will see in the following chapters that adding community as a feature in an ML model can drastically increase model performance.

Summary

In this chapter, you have learned the basic principles of the Neo4j GDS library 2.x. You have been able to create projected graphs, configuring included nodes, relationships, and properties with native graph projection. You have also learned how to generate properties or relationships on the fly using Cypher projections. In the second section, you have run your first GDS algorithm—the degree algorithm—and got familiar with the `stream`, `write`, and `mutate` algorithm modes. You have also been made aware of the algorithm configuration, especially regarding relationship orientation.

Once GDS had no more secrets to you, we started using other types of algorithms—namely, community detection algorithms. We studied a few of them and learned about their differences and what they can teach us about our graph.

In the next chapter, we will learn how to use another powerful tool of the Neo4j universe: Neo4j Bloom, yet another graph application. Bloom is designed to make graph visualization accessible to everyone, and, as we all know, with great visualization comes great insights.

Further reading

To investigate further the topics covered in this chapter, you can check the following resources:

- The GDS manual: `https://neo4j.com/docs/graph-data-science/current/`.
- *Hands-On Graph Analytics with Neo4j*, my previous book, gives much more detail about each centrality and community detection algorithm, including example implementations to better understand what they are doing.
- *NEuler - the Graph Data Science Playground*: An application, similar to Neo4j Desktop or `neodash`, to run graph algorithms from GDS. At the time of writing, it was not yet updated to work with Neo4j 5, but it's worth keeping an eye on it since it can be very useful for investigations: `https://github.com/neo4j-devtools/neuler`.

5

Visualizing Graph Data

Graphs are special objects. Unlike images, there is no simple way to visualize them. The preceding chapters have demonstrated how we can extract information from a graph dataset: node importance using centrality metrics (for example, degree) or node clusters with community detection algorithms (for example, the Louvain algorithm). We have also already used some tools to visualize the content of our graph: *neodash* to draw charts from data stored in Neo4j, and Neo4j Browser, which is able to draw a graph with nodes and relationships in a dynamic way. Neo4j Browser is very convenient to see the result of a Cypher query, but it is not intended for data analysis visualization. Typically, it does not let us configure node color based on a node property.

In this chapter, we will focus on graph data visualization. We will first learn why it is challenging and what the graph visualization techniques are. We will first create static but customizable images of a graph in Python using the `networkx` and `matplotlib` packages. In the second part, we will install and use **Neo4j Bloom**, a powerful graph application created by the Neo4j team containing many essential features to deal with graphs in a no-code manner. Finally, we will draw a graph visualization using a famous graph visualization tool, **Gephi**, a must-have tool in a graph analyst's toolbox, and learn how to interface it with Neo4j.

In this chapter, we're going to cover the following main topics:

- The complexity of graph data visualization
- Visualizing a small graph with `networkx` and `matplotlib`
- Discovering the Neo4j Bloom graph application
- Visualizing large graphs with Gephi

Technical requirements

In order to be able to reproduce the examples given in this chapter, you'll need the following tools:

- Neo4j 5.x installed on your computer (see the installation instructions in *Chapter 1, Introducing and Installing Neo4j*) with the following plugins installed:
  - **Awesome Procedures on Cypher** (**APOC**)
  - GDS plugin (*version >= 2.2*)
- A Python environment with Jupyter to run notebooks
- An internet connection to download the plugins and the datasets
- Any code listed in the book will be available in the associated GitHub repository (`https://github.com/PacktPublishing/Graph-Data-Science-with-Neo4j`) in the corresponding chapter folder

The complexity of graph data visualization

In order for us to understand why graph visualization is so challenging, we are first going to investigate the easiest networks that can be visualized: physical networks.

Physical networks

By physical networks, I mean networks whose nodes (and sometimes edges) have fixed spatial positions (coordinates). That includes the following:

- **Road networks**: Street intersections (nodes) have spatial coordinates (latitude and longitude). Edges (the roads themselves) also have a shape or geometry (linestring) that can be stored using a geospatial data format (`shapefile` or `GeoJSON`, for instance) and drawn on a map.
- **Public transport networks**: Nodes are bus/train stops with defined positions; edges are the bus paths between these stops.
- **Electric network**: We can imagine this as containing nodes with different types (power station, transformer, consumer, etc.). Each of them also has a precise location, and the distance between them can be measured.
- **Electric circuit**: On a printed circuit board, such as the one making the motherboardcard in your computer, electronic components are welded into a board encapsulating conductive paths to connect these components. Here, nodes won't have latitude and longitude, but they can be given coordinates in the board referential.

In each of these cases, nodes have coordinate properties that can be used to draw them on a 2D frame. The following screenshot is such an example, showing the street network in Paris (France), extracted from **OpenStreetMap** (**OSM**):

Figure 5.1 – Paris road network from OSM

If you fall into this case, lucky you!

As you can imagine, things become much more complex when we can't find such fixed coordinates for nodes. That's the general case we are going to discuss next.

General case

In the general case, nodes do not have coordinate properties we can use to plot them. So, any graph visualization will have to find out such coordinates, depending on some constraints.

We've all seen graph visualizations such as this:

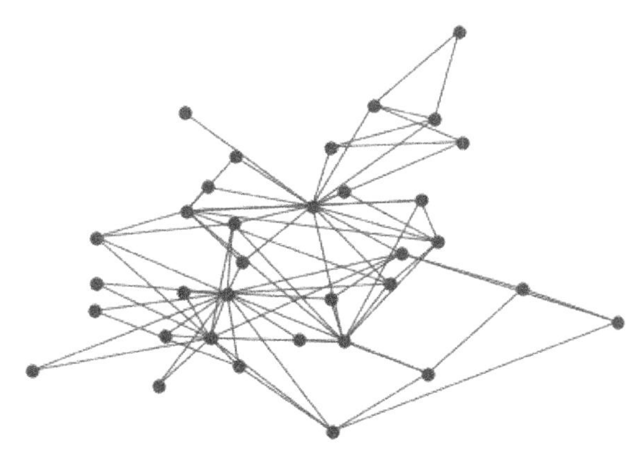

Figure 5.2 – Graph data visualization

But we are not always aware that the previous graph visualization can also be represented in many different ways. A few examples are given in the following screenshot:

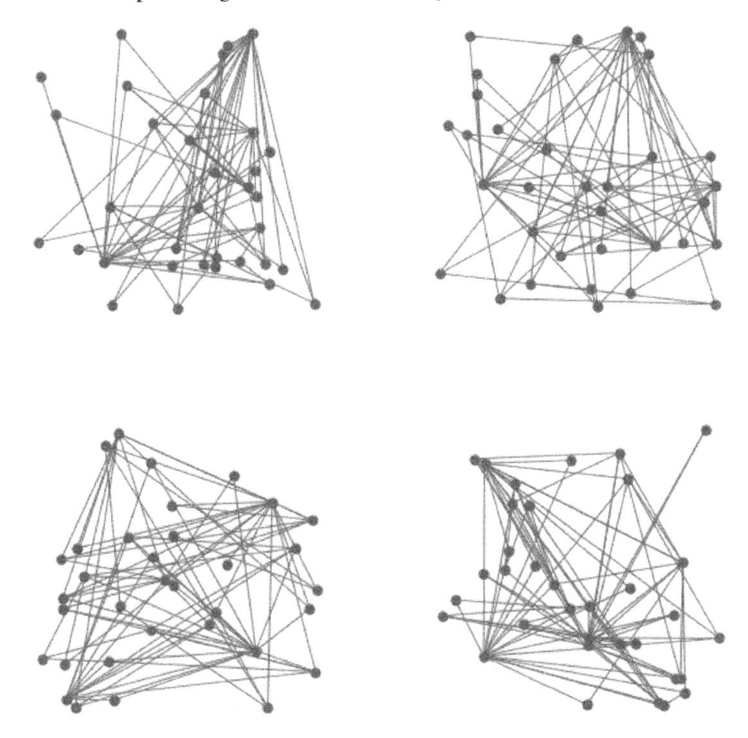

Figure 5.3 – Random graph layouts

All five visualizations (in *Figures 5.2* and *5.3*) represent the exact same graph, in the sense that they contain the exact same number of vertices and the exact same edges between these vertices. But the position of these vertices on the 2D plane is totally different, making them look like they are different objects.

> **Note**
>
> That means that graph visualization is not unique. Viewing (or not viewing) a pattern in a given graph visualization is not proof that this pattern actually exists (or does not exist).

In the preceding screenshots, I have used random numbers to generate node positions, but other techniques exist in order to try to extract some visual information from the graph. In the following section, we will first learn how to draw these kinds of visualizations for small graphs and start exploring several graph layouts.

Visualizing a small graph with networkx and matplotlib

When the graph is small enough, such as the ones represented in the previous screenshots (*Figure 5.2* and *5.3*), it can be convenient to visualize them using the `matplotlib` plotting library. In this section, we are going to reproduce the visualizations displayed previously.

When dealing with graphs in Python, fortunately, we do not have to create our own data structure and implement our algorithms. As with many other tasks, we can just `pip install` a package developed by the fantastic open source community around Python. For graphs, the most used package is called `networkx`. Let's go ahead and go through our next Jupyter notebook.

Visualizing a graph with known coordinates

In this section, we are going to draw a graph representing a part of the road network around the Colosseum in Rome. This data was extracted using the `osmnx` package, but we are not going to detail its extraction process here, even if `osmnx` makes it very simple. If you are interested, refer to the `Geospatial_Network_Creation` notebook, where you'll be able to extract data for other places if you're interested. The code in this section is available in the `Spatial_Graph_visualization_with_networkx_and_matplotlib` notebook. Let's go over it:

1. First, install the requirements if not already installed:

   ```
   !pip install  networkx==3.0 matplotlib==3.6.3
   ```

   You can also install from Jupyter or from a Terminal (in the same virtual env used by Jupyter if you are using one) as follows:

   ```
   $ pip install  "networkx==3.0" "matplotlib==3.6.3"
   ```

2. In your notebook, import the necessary packages:

   ```
   import networkx as nx
   ```

3. Read the graph data (stored in *GraphML* format) and create a *networkx* graph object:

   ```
   G = nx.read_graphml(
         "../data/osm_street_network_rome.graphml"
   )
   ```

   Update the path to the data file if you are not using the same folder structure.

   The graph we've just loaded contains 1,116 nodes with latitude and longitude properties and 2,115 edges.

> **Naming convention**
>
> It is a `networkx` convention to name the graph object with a capital *G* letter.

4. Extract the node position from the node properties:

```
pos = {
    node_id: (
        node_data["longitude"],
        node_data["latitude"]
    )
    for node_id, node_data in G.nodes(data=True)
}
```

Here, we are using Python list comprehension to extract longitude and latitude from the node's data (accessed through the `G.nodes(data=True)` method).

5. Finally, we can draw the graph using our computed positions for nodes, with the following line of code:

```
nx.draw(G, pos=pos, node_size=10)
```

Note about spatial coordinates

When extracting latitude and longitude coordinates, we used the longitude as the first coordinate and the latitude as the second. This is the way to draw a network as close to reality as possible. You can understand this if you imagine moving along the equator (the horizontal—hence *x*—axis): your latitude doesn't change (it's always 0), but your longitude does as you cross different meridians. So, in spatial data, the *x* axis corresponds to longitudes and the *y* axis to latitudes.

The result of the preceding `nx.draw` function is the network represented in the following screenshot, where we can see nodes in blue (street intersections) and streets connecting these nodes in black:

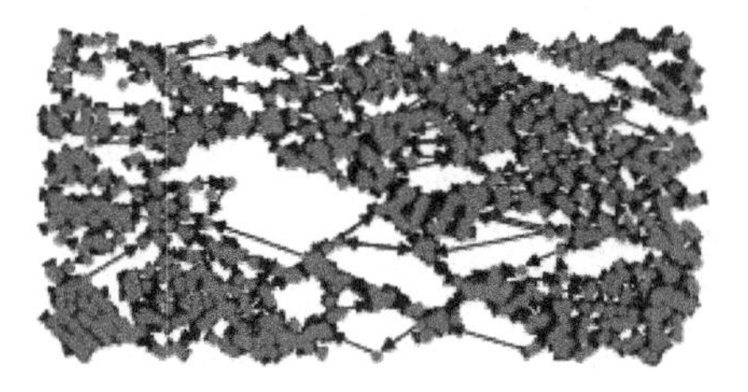

Figure 5.4 – Street network visualization

As I was mentioning earlier, this type of graph visualization is the easiest one since we know where each node must be drawn. But this is not the case in most situations. If your nodes are users or products instead of street intersections, what is the *position* to use for them? In the next section, and in the rest of this chapter, we will only deal with such graphs, without any *a priori* known node position.

Visualizing a graph with unknown coordinates

In the rest of this section, we are going to use a predefined graph called **Zachary's Karate Club** (ZKC). It has been studied at great length in graph theory and is hence included by default in `networkx` (similarly to scikit-learn's ability to easily create some toy datasets such as the IRIS or digits datasets).

Creating a ZKC graph with networkx

The code for this section can be found in the `ZKC_Graph_Visualization_with_networkx_and_matplotlib` notebook.

Creating a ZKC graph is a matter of one line of code:

```
G = nx.karate_club_graph()
```

This graph is an undirected graph (of type `nx.Graph`) that contains 34 nodes and 78 edges. This information can be checked with the following code:

```
type(G), G.number_of_nodes(), G.number_of_edges()
```

We can further check the content of the graph by inspecting the content of the first two nodes:

```
list(G.nodes(data=True))[:2]
```

And we can do the same for the first two edges:

```
list(G.edges(data=True))[:2]
```

The results for the nodes are shown here:

```
# nodes
[(0, {'club': 'Mr. Hi'}),
 (1, {'club': 'Mr. Hi'})]
```

And here are the results for the edges:

```
# edges
[(0, 1, {'weight': 4}),
 (0, 2, {'weight': 5})]
```

Here is the information we can extract from these results:

- Nodes have the following attributes:
  - Integer IDs (0 and 1, displayed in the preceding code snippets).

    - A (string) property called `club`. We can check the content of this property with the following code:

    ```
    G.nodes.data("club")
    ```

    - It shows that they are only two possible values for the club property: *"Mr. Hi"* or *"Officer"*.

- Edges have the following attributes:

    - They are represented by a set of two node IDs. The preceding code sample shows edges from node 0 to 1 and from node 0 to 2.
    - A property named `weight`, containing a number.

Let's go ahead and draw a ZKC graph.

Drawing a graph with unknown coordinates

As we discussed in the first section of this chapter, the challenge here is to find coordinates for nodes, since they do not have any. The easiest solution is to define these coordinates randomly, and that's what the `nx.random_layout` function does.

Here is how we draw a graph using `networkx` (which, internally, relies on `matplotlib`).

Assign some coordinates to nodes:

```
pos = nx.random_layout(G, seed=42)
```

Draw the graph using these coordinates:

```
nx.draw(G, pos=pos)
```

Checking the content of the `pos` variable, you'll see that it is a mapping between node IDs and a 2D NumPy array, representing the (randomly) computed coordinates. The plotted graph looks like the one displayed in the following screenshot:

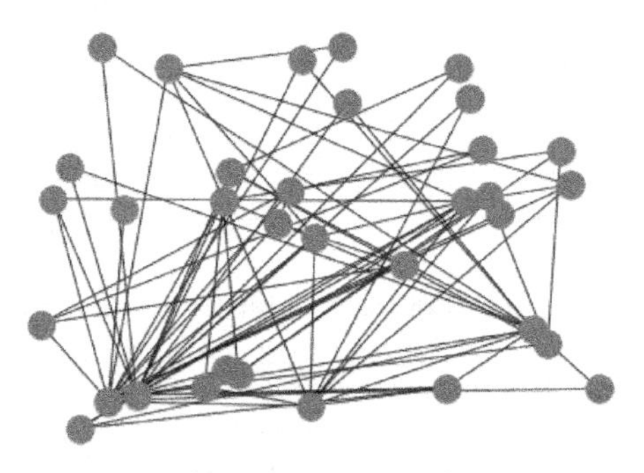

Figure 5.5 – ZKC graph with a random layout

Randomness

The `seed` parameter is used in order to obtain reproducible results. If you run the cell containing the random layout creation, you'll obtain the exact same results. If you want to try another layout, you can change the seed, or set it to `None`, which will let the `random` module do its job.

Fortunately, randomness is not the only way to define node positions, and some algorithms have been designed to try to render graphs in a more readable manner. We are not going to detail the internal functioning of these algorithms but will instead draw the same graph with different layouts in order to compare them.

Understanding graph layouts

The `networkx.draw` package contains several built-in layouts for graphs. Here is how to use them with the corresponding graph visualization:

- Circular layout:

```
pos = nx.circular_layout(G)
nx.draw(G, pos=pos)
```

  This code produces the following output:

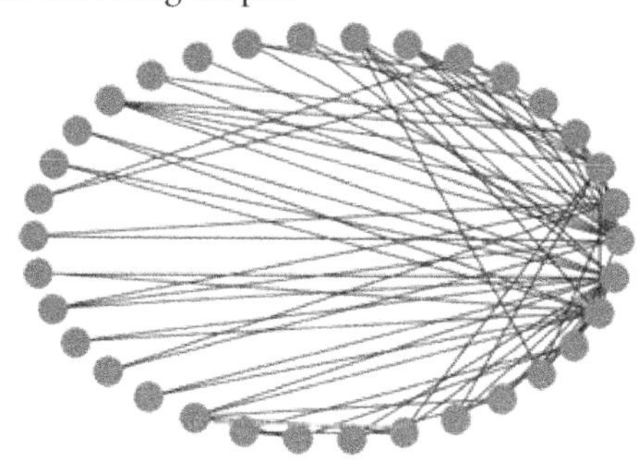

Figure 5.6 – ZKC graph with a circular layout

- Kamada-Kawai layout:

```
pos = nx.kamada_kawai_layout(G)
nx.draw(G, pos=pos)
```

This is illustrated in the following screenshot:

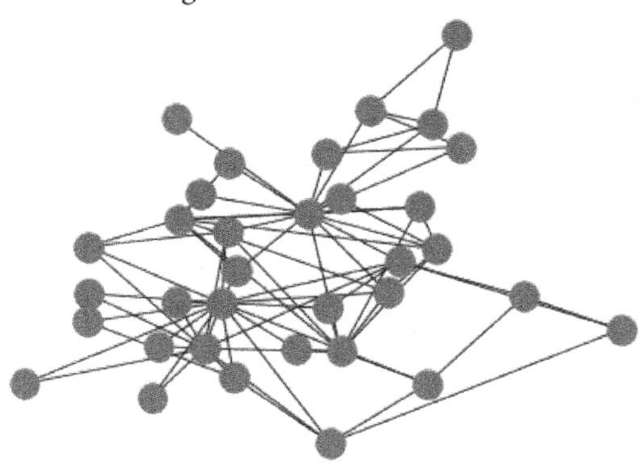

Figure 5.7 – ZKC graph with a Kamada-Kawai layout

- Spring layout:

```
pos = nx.spring_layout(G)
nx.draw(G, pos=pos)
```

The spring layout is represented in the following screenshot:

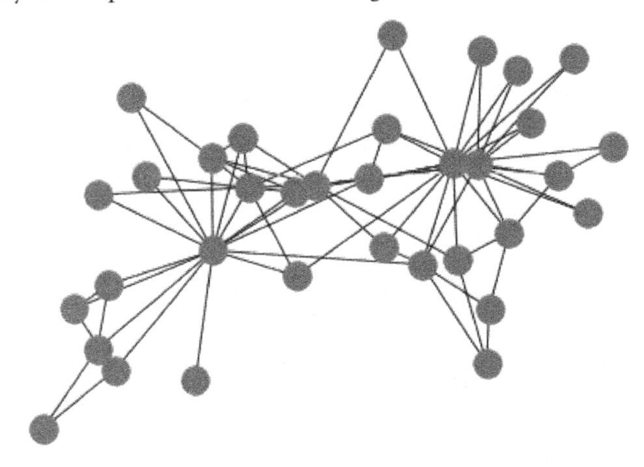

Figure 5.8 – ZKC graph with a spring layout

All of the three preceding screenshots represent the same graph: the exact same number of nodes and the exact same edges. The only difference is the position of each node in the 2D plane where our graph is drawn. It is important to realize that none of the preceding representations is true or false, but some of them might be more helpful in some situations.

In the next section, we are going to see how to customize node and edge display properties—typically, their color and size.

Configuring object display

For once, let's start with the code:

```
nx.draw_spring(
    G,
    node_size=150,
    node_color=[
        "black" if node["club"] == "Mr. Hi" else "grey"
        for _, node in G.nodes(data=True)
    ],
    width=[e["weight"]/2. for _, _, e in G.edges(data=True)]
)
```

And now, let's understand what it is doing:

- It uses the `nx.draw_spring` shortcut function, which is equivalent to this:

  ```
  pos = nx.spring_layout(G)
  nx.draw(G, pos=pos)
  ```

- It decreases the node size to 150 (the default is 300).
- It sets the node color depending on the node's `club` property: if this property's value is *Mr. Hi*, then the node will be displayed in black; otherwise, it will be gray.
- It sets the edge width proportionally to the edge's `weight` property; the higher the weight, the wider the edge's line.

The graph obtained by running the preceding code is reproduced in the following screenshot:

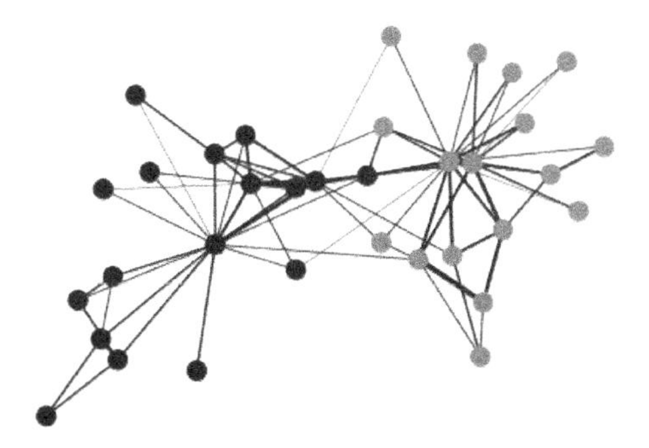

Figure 5.9 – Customizing node and edge display properties (node color and edge width)

As you can see from the code, node and edge color and width can be parametrized in two ways:

- By providing a single value that will be identical for all objects (for example, the node size in our example)
- By providing a list where item number `k` contains the display property for node `k` (that's what we have done for node color and edge width)

`networkx` is a very nice package for managing graphs in Python. It does handle graph objects (undirected or directed) and graph visualization, but also contains implementations for graph algorithms we have studied in the preceding chapters of this book.

But when it comes to large graph visualization, static views—such as the ones created by `matplotlib`—can quickly become too busy and unreadable. That's where other software enters the game and can be really useful. In the coming section, we are going to discover a Neo4j-specific graph management software: Neo4j Bloom.

Discovering the Neo4j Bloom graph application

Neo4j Bloom is distributed by Neo4j and is a professional-looking application to deal with data specifically stored in a Neo4j database. In this section, we are going to discover its features and use its ability to configure node display color based on a property to visualize graph communities.

What is Bloom?

What can Bloom do for us? It can do all of the following:

- **Graph querying**:
    - By label
    - By path
    - With a parametrized Cypher query
- **Graph visualization**:
    - Filtered nodes, customized sizes, color, displayed properties, and so on
- **Graph editing**:
    - Adding nodes and relationships
    - Adding/editing properties

- **Graph exploration**:
    - Shortest path
    - GDS integration

In this chapter, I'll leave apart the graph editing functionalities to focus on graph querying and graph data visualization features, taking a small tour of the GDS integration.

Bloom installation

If you are using Neo4j Desktop, chances are that it is already installed! Check it here in the Neo4j Desktop application:

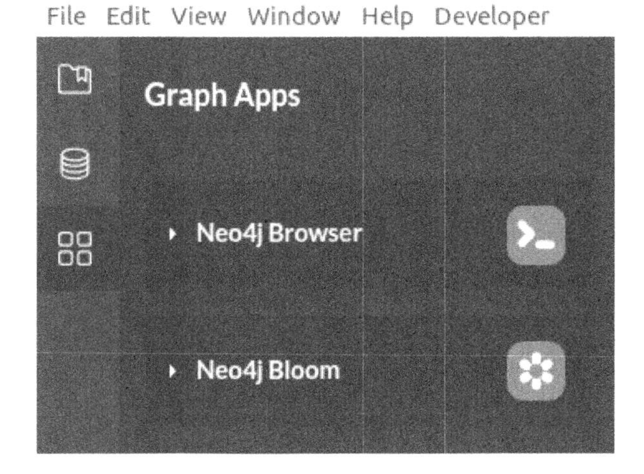

Figure 5.10 – Checking that Neo4j Bloom is an available graph application

If you can't see **Neo4j Bloom**, you can easily install it with the same instructions we used earlier in this book to install neodash or NEuler (refer to the *Visualizing the degree distribution with Neodash* section of *Chapter 3, Characterizing a Graph Dataset*).

Selecting data with Neo4j Bloom

In this section, we are going to use the Netflix dataset we have been exploring since the beginning of this book. As a reminder, it contains `Person` nodes representing actors and directors of movies and TV shows available on the streaming platform. We have already performed component analysis and added the `MainComponent` label to nodes in the biggest connected component. This component still contains around 37,000 nodes, so we have also run the Louvain algorithm on nodes in the main component to identify sub-structures and communities. More than 100 communities have been identified by this algorithm, with a population ranging from 5 to 4,940 nodes.

In order to create a visualization in which we can actually see something, we are going to select only the top 5 communities. It is still 14,691 nodes, higher than the maximum number of nodes Bloom can show at the same time (limited to 10,000).

We will, then, start by adding a specific label to the nodes we want to show (those in the top 5 biggest Louvain communities), and then use Bloom to display them:

1. If you have not followed along from previous chapters, you can load a database dump. Check the README file in the code files corresponding to this chapter for instructions.
2. Identify the top 5 biggest Louvain communities and assign a specific label to nodes in these communities:

```
MATCH (node:MainComponent)
WITH node.louvain as louvain, count(node) as ct
WITH louvain, ct
ORDER BY ct DESC LIMIT 5
MATCH (mc:MainComponent {louvain: louvain})
SET mc:MainLouvainTop5
```

We are now ready to start Bloom and visualize some data. Proceed as follows:

1. Start **Neo4j Bloom** from Neo4j Desktop, as follows:

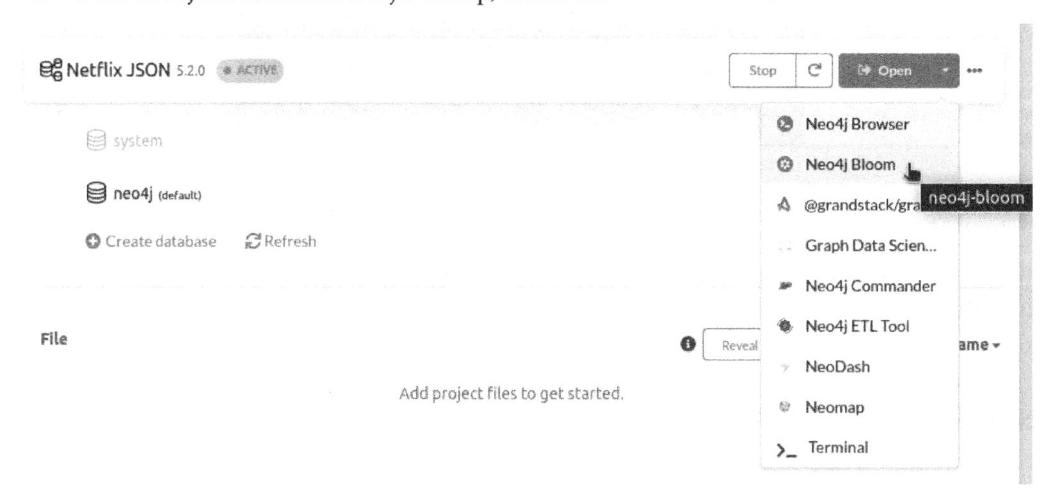

Figure 5.11 – Starting Bloom from Neo4j Desktop

It will automatically connect to your active graph.

2. The default view contains a node selector in the upper-left corner of the screen. Let's go ahead and select all **MainLouvainTop5** nodes connected to another node of the same label, as illustrated in the following screenshot:

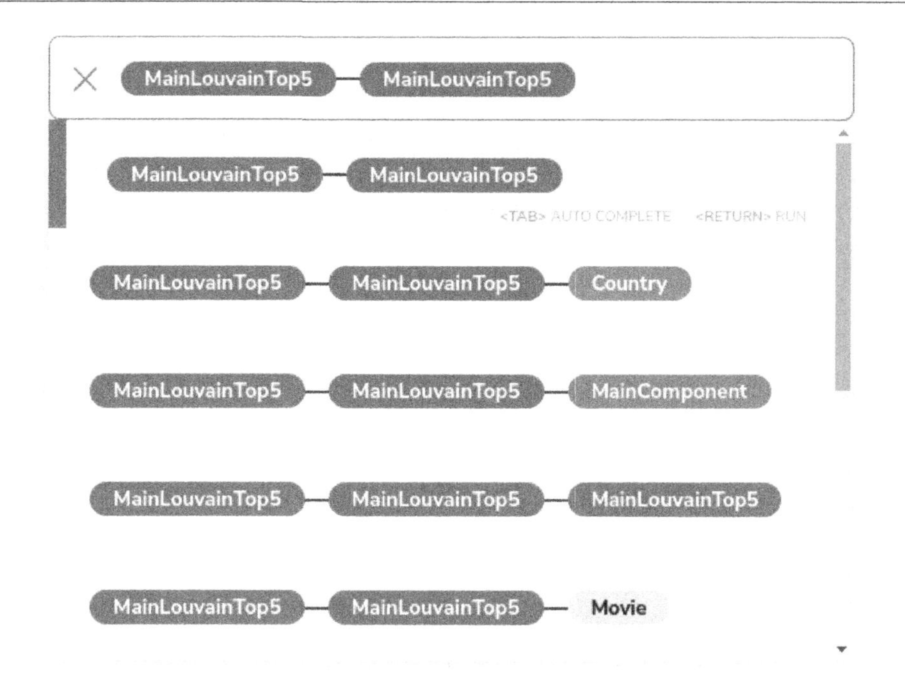

Figure 5.12 – Neo4j Bloom node selection

3. After hitting *Enter*, you'll see something like the following view:

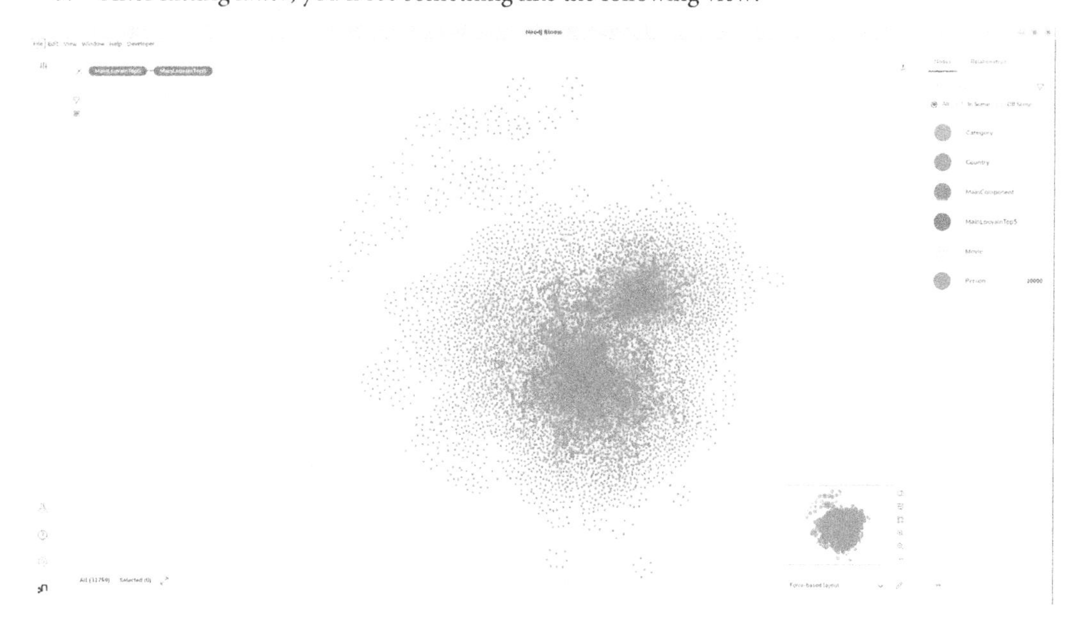

Figure 5.13 – Neo4j Bloom default view

4. All nodes with a **Person** label are displayed in violet. If you double-click a node, you'll be able to see its properties (which you can edit) and its neighbors (see the following screenshot):

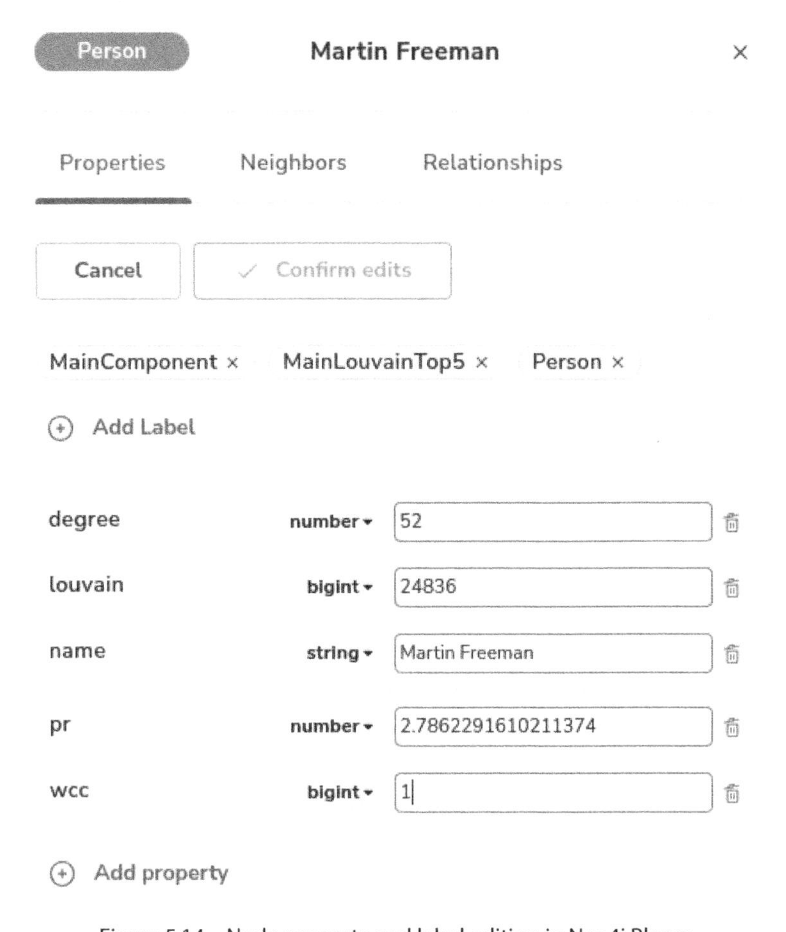

Figure 5.14 – Node property and label editing in Neo4j Bloom

In the following section, we are going to use another feature of Bloom: its ability to tune the visualization to try to see whether the communities found by the Louvain algorithm earlier can be visualized.

Configuring the scene in Bloom

In order to configure the node colors, we are going to use the right configuration bar. Follow these steps:

1. Click on the blue circle next to the **Person** label.
2. In the opened panel, click on **Rule-based**.

3. From the displayed dropdown, you can select the node property to use to differentiate nodes. Select **louvain**.
4. Check the **Unique values** checkbox below the dropdown.
5. Finally, click on **Apply color**.

   The resulting view should look like the following screenshot:

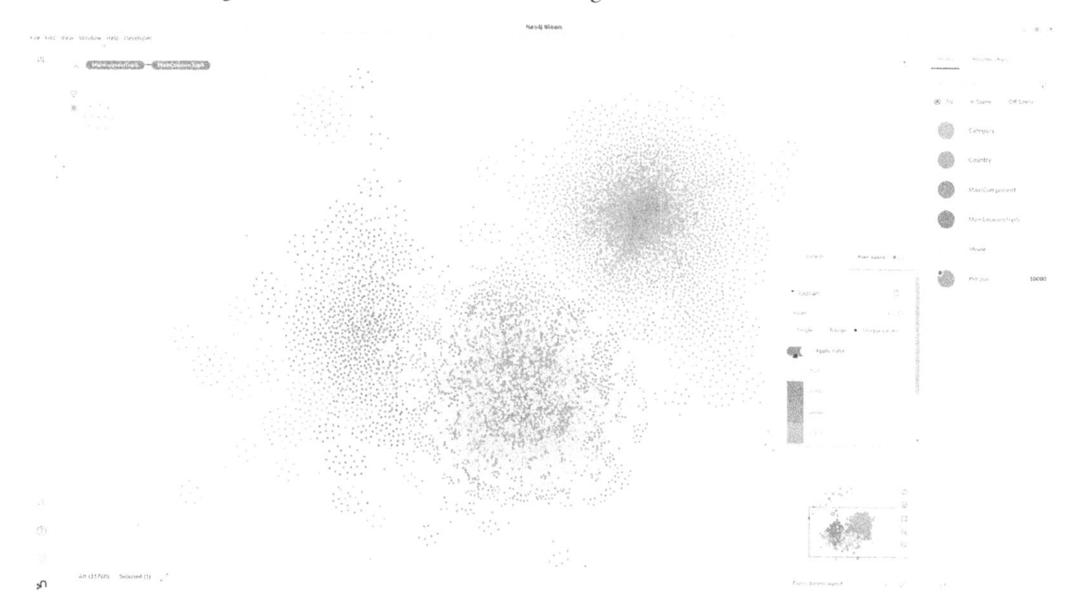

Figure 5.15 – Neo4j Bloom custom node colors

The view is much nicer, and communities are immediately made more visible. The nodes in orange all belong to the same Louvain community as the nodes in red or violet.

> **Maximum number of nodes**
>
> By default, the maximum number of nodes in Bloom is set to 8,000 (the number of nodes that will be fetched from the database). It can be increased up to 10,000 in the **Settings** panel.

Last but not least, let's mention that Bloom also integrates some functionalities from the GDS library. By enabling the **Data science** support from the upper-left part of your screen, you can run algorithms and use them to configure graph rendering (see the following screenshot):

Figure 5.16 – Neo4j Bloom data science mode

Neo4j Bloom already allows us to draw dynamic graph visualizations, where nodes can be moved around and properties displayed with a simple click. But it does not offer much control over the graph layout itself.

When dealing with graph **machine learning** (**ML**), another important software we need to have in our toolbox is Gephi.

Visualizing large graphs with Gephi

Gephi is a powerful open source graph visualization software, able to deal with very large graphs. In the following sections, we are going to install Gephi and the required plugins, set up our Neo4j database, and draw a graph using this software.

Installing Gephi and its required plugin

In order to install Gephi and the plugins we need to connect it with Neo4j, follow these steps:

1. Download Gephi from `https://gephi.org/`.
2. To start it, follow your OS-specific instructions. For Linux, the following commands should work (make the necessary changes depending on the version of Gephi you downloaded):

```
cd Downloads/
tar xzvf gephi-0.9.7-linux-x64.tar.gz
cd gephi-0.9.7/bin
./gephi
```

3. Install the streaming plugin from the Gephi UI.
4. Open the plugins wizard from the **Tools | Plugins** menu.
5. Go to the **Available Plugins** tab.
6. Search for the **Graph Streaming** plugin in the list and select it.
7. Click **Install**.

   The following screenshot shows the **Graph Streaming** plugin installation:

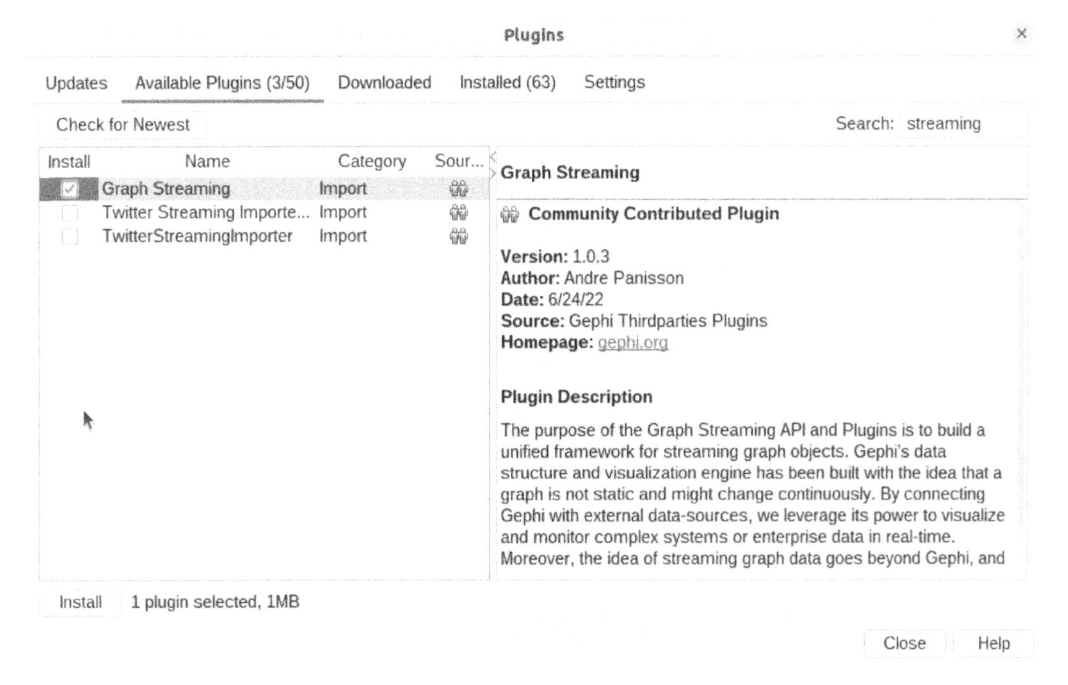

Figure 5.17 – Gephi Graph Streaming plugin installation

   After the plugin is installed, we need to configure it.

8. Show the **Window | Streaming** streaming pane.
9. Optionally, disable SSL—click **Settings** in the streaming pane and uncheck **Use SSL** (see the following screenshot):

Figure 5.18 – Gephi Graph Streaming settings

10. Finally, start the server—right-click on **Master Server** and click **Start**:

Figure 5.19 – Gephi Graph Streaming: starting the master server

Gephi is now ready to be used with Neo4j. The other part of the configuration has to be performed directly from Neo4j, using the famous APOC plugin.

Using APOC Extended to synchronize Neo4j and Gephi

Since Neo4j 5, the APOC plugin has been split into two parts:

- **Core**: The procedures we have used so far to load JSON files or query HTTP APIs, for instance.
- **Extended**: Contains even more functionalities not useful to everyone. Integrations with other tools, such as Gephi, are part of this extended plugin.

The Gephi-related procedure is part of the extended plugin, which we are going to install now. It is not yet possible to install it as easily as the core version through the Neo4j Desktop application, but the installation remains quite straightforward. Follow these steps:

1. Visit the release page of the plugin on GitHub: `https://github.com/neo4j-contrib/neo4j-apoc-procedures/releases`.
2. Pick the release matching your Neo4j version. At least the first two digits have to match. For instance, if using Neo4j 5.2.1, you can use APOC Extended version 5.2.0.
3. Download the `apoc-<VERSION>-extended.jar` JAR file.
4. Then, from Neo4j Desktop, open the **Plugins** folder, as illustrated in the following screenshot, starting from the **...** symbol next to your active graph:

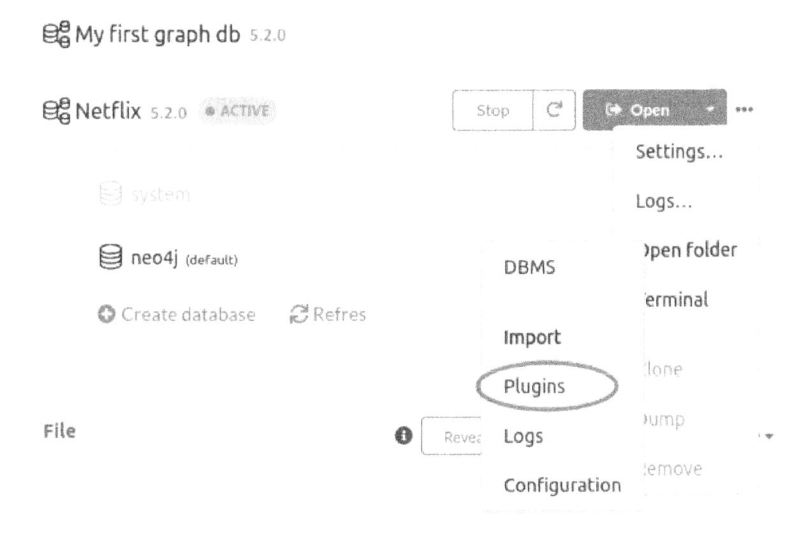

Figure 5.20 – Opening the Plugins folder

5. Copy the JAR file you've just downloaded into this `Plugins` folder.
6. Restart your database.
7. Check the new procedures are available:

```
SHOW PROCEDURES YIELD name
WHERE name STARTS WITH 'apoc.gephi'
RETURN name
```

This should return the `apoc.gephi.add` procedure we are going to use right away.

This procedure is used to send data to the Gephi streaming plugin. Run the following query in Neo4j Browser:

```
MATCH path = (:MainLouvainTop5)-[:KNOWS]-
>(:MainLouvainTop5)
WITH path LIMIT 50000
CALL apoc.gephi.add(null,'workspace1',path, "",
['louvain'])
YIELD nodes, relationships, time
RETURN nodes, relationships, time
```

Let's understand this query:

I. First, we select all paths from one person from `MainComponent` to another through a `KNOWS` relationship and store each of these paths in the `path` variable.

II. Then, we instruct APOC to send data to Gephi, as follows:

- Use the default Gephi URL (`http://localhost:8080`) (hence `null` as the first argument).
- Use `workspace1` from Gephi (the default workspace created when starting the application).
- Send the `path` variable, including start and end nodes and the relationship between them.
- Do not include any weight property for edges (the empty string parameter).
- But send to Gephi the `louvain` node properties.

That's it! Our Neo4j graph is not accessible from Gephi. Go back to the Gephi window for the following sections.

> **Note**
>
> The data is streamed from Neo4j to Gephi. Later, you will see your graph growing on the screen as new nodes are received. It can take a few minutes for the 50,000 rows we are fetching from Neo4j to be ready in Gephi—wait for it.

Configuring the view in Gephi

Similarly to what has been done with Neo4j Bloom, we are going to configure node and edge colors. We are also going to deal with the graph layout in order to create a nice visualization of our graph.

The initial view in Gephi looks like this:

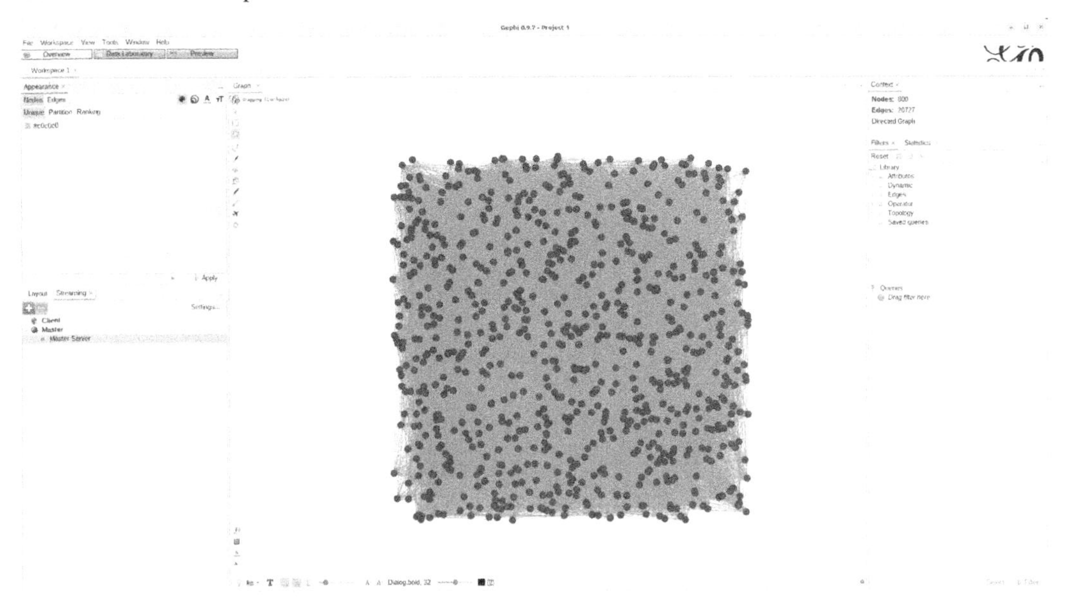

Figure 5.21 – Gephi default view: colors may differ from yours

Let's tune it to better understand our graph dataset.

Setting the node color

In order to set the node color, follow these steps (refer to the following screenshot to see where to find each option on your screen):

1. From the left panel, select the **Nodes** tab.
2. Go to the **Partition** tab.
3. Select the **louvain** property from the drop-down menu.
4. Click **Apply**.

The resulting view is similar to the one displayed in the following screenshot:

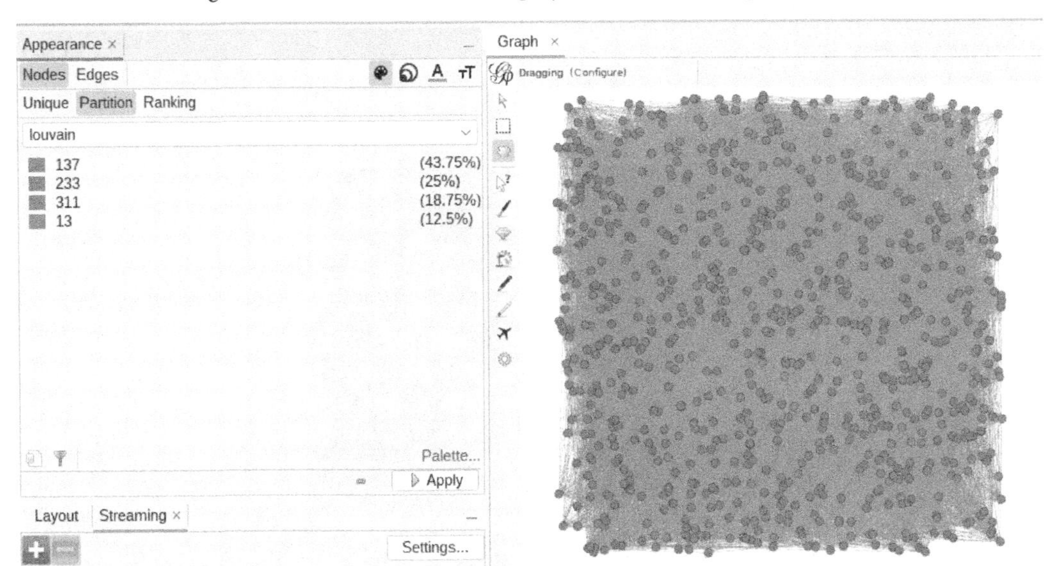

Figure 5.22 – Gephi custom node color

It is not yet super helpful. We are first going to reduce the visual importance of edges by changing their color.

Setting the edge color

In order to configure the edge color, the steps are almost identical to the nodes case:

1. From the left panel, select the **Edges** tab.
2. Go to the **Unique** tab (default).
3. Select a color.
4. Click **Apply**.

We end up with the following graph visualization, which starts to be a bit clearer:

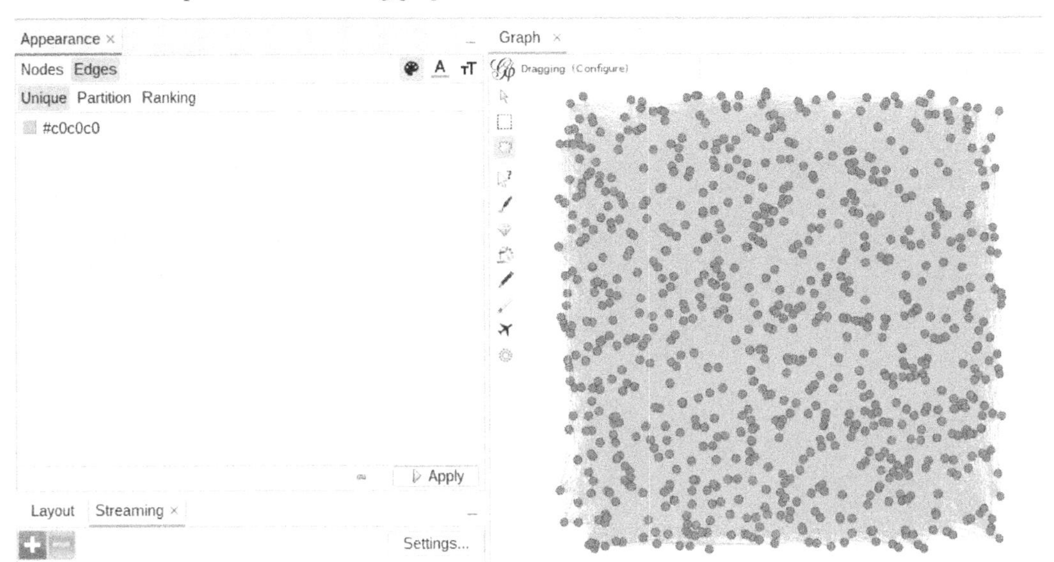

Figure 5.23 – Gephi custom edge color

But compared to the Neo4j Bloom view, we still can't *see* the communities. In order to make them appear, we'll have to deal with the graph layout, similar to what we did with `networkx` earlier in this chapter.

Setting the graph layout

Let's go ahead and change the node positions on the screen. To do so, go to the **Layout** tab at the bottom left of your screen (close to the **Streaming** tab), as illustrated in the following screenshot:

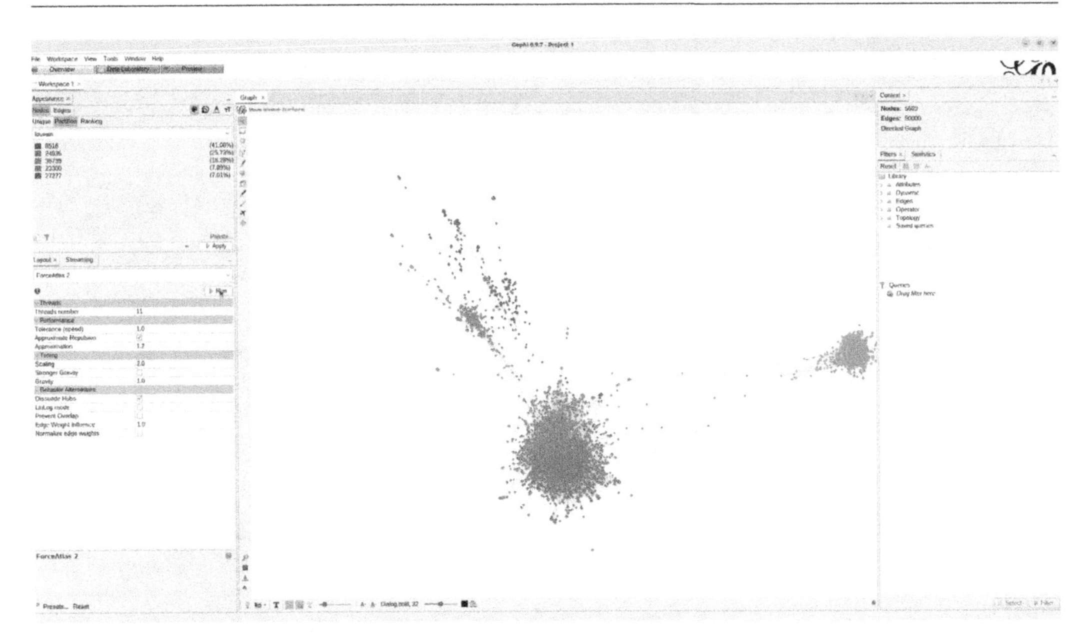

Figure 5.24 – Gephi graph layout

From there, select the **ForceAtlas 2** layout, and click **Run**. Almost immediately, nodes will re-organize and communities appear, since we have already configured the node colors based on the partition they belong to.

> **Tip**
>
> Use your mouse's right button to move the graph on the screen.

That's it for our quick tour of Gephi. In a few minutes only, we have been able to connect it to Neo4j and visualize a graph in a very nice way. There are many more configuration options in Gephi, such as node size and edge width. You're all set up to explore its capabilities on your own.

Summary

In this chapter, we've explored a few helpful tools for graph data visualization. First, `networkx` has helped us visualize relatively small graphs in a Jupyter notebook. We have explored the challenge of graph data visualization and learned about graph layout. In the second part, we have used another great tool—part of the Neo4j ecosystem—called Neo4j Bloom. It has many features allowing to deal with graph data stored in Neo4j without writing any Cypher query. We have focused on how to customize the appearance of the graph, choosing the node's color and size.

Finally, we have discovered a very powerful tool we have to know about when dealing with GDS: Gephi. Here, again, we have focused on node appearance configuration.

In all cases, you are highly encouraged to dig deeper into these tools by yourself, using your own data and/or exploring the features we can't talk about in this book (unless we double its length, but then nobody would read it!).

This chapter closes the part of this book dedicated to graph dataset characterization and visualization. You are now able to use Cypher, APOC, the GDS library, and a few graph applications (neodash, Bloom) to find information about your graph data. You also know how to retrieve data from Neo4j using Python and the Neo4j Python driver, which is crucial for using graph data in a data science pipeline.

In *Part 3: Making Predictions on a Graph*, we will start to actually make predictions on a graph! In the next chapter, we will start by using a well-known Python library for ML, scikit-learn, and leverage what we have learned so far about graph algorithms to build an ML model that takes into account the graph structure of our dataset.

Further reading

To dig deeper into the concepts covered in this chapter, you can start with the following resources:

- NetworkX documentation about graph visualization: `https://networkx.org/documentation/stable/reference/drawing.html`
- Neo4j Bloom demo: `https://neo4j.com/developer/neo4j-bloom/`
- Neo4j Bloom reference guide: `https://neo4j.com/docs/bloom-user-guide/current/`
- Gephi documentation: `https://gephi.org/users/`
- A nice resource about graph layouts if you want to know more about how they are computed is found in the *2014 CRC Press Handbook of Graph Drawing and Visualization* by *Roberto Tamassia*. You can check *Chapter 11* of *Simultaneous embedding of planar graphs* by *Bläsius, et al,* for more.

Exercises

To practice what you have learned in this chapter, you can use the following ideas to explore your data:

1. Try to build the street network of your own location. For this, you will need to find the central location coordinates (which you can do using Google Maps, for instance) and update the `Geospatial_Network_Creation` notebook.
2. In Neo4j Bloom, use the **Filter** toolbox to visualize only nodes in the bigger Louvain community (use Cypher to find out the ID of the biggest community).
3. Still in Bloom, configure the node color to be a function of its degree (the value stored in the `degree` property).

Part 3 – Making Predictions on a Graph

After building a graph and learning how to characterize and visualize it, it's time to learn about techniques to make predictions from graph data. You will start by using a well-known Python library, namely scikit-learn, and extract data from Neo4j to build a model. Then, you will learn about node embedding algorithms that are built to automatically create node features based on the graph structure. You will then use these embeddings to build node classification and link prediction pipelines, without the need for a third-party library since everything will be managed by the **Graph Data Science** (**GDS**) library.

Finally, in the last chapter, you will build a GDS extension and write your own graph algorithm that behaves in the same way as all built-in GDS procedures.

This part includes the following chapters:

6

Building a Machine Learning Model with Graph Features

So far, we have explored how to build and understand a graph dataset. We have computed some key metrics, such as the degree distribution. Using unsupervised algorithms, such as community detection, we have also better-identified graph structures. Finally, the graph visualization tools have enabled us to see the content of the dataset, and visually identify some aspects of the graph. Now, it is time to start applying this knowledge to build a **machine learning** (**ML**) model. In this chapter, we will introduce the Python client for the **Graph Data Science** (**GDS**) library, by allowing it to run graph algorithms directly from Python, without writing any Cypher. After computing and extracting our features from Neo4j, we will build a scikit-learn pipeline to train a model and make predictions.

In this chapter, we're going to cover the following main topics:

- Introducing the GDS Python client
- Running GDS algorithms from Python and extracting data in a dataframe
- Using features from graph algorithms in a scikit-learn pipeline

Technical requirements

In order to be able to reproduce the examples given in this chapter, you'll need the following tools:

- Neo4j 5.x installed on your computer (see the installation instructions in the first chapter)
- The GDS plugin (*version >= 2.2*)
- A Python environment with Jupyter to run notebooks
- An internet connection to download the plugins and the datasets
- Any code listed in the book will be available in the associated GitHub repository (`https://github.com/PacktPublishing/Graph-Data-Science-with-Neo4j`) in the corresponding chapter folder

Introducing the GDS Python client

The previous chapters introduced the Neo4j GDS library. There, you discovered the concepts of projected graphs and the procedures to run specific graph algorithms from Cypher. If you do not have direct access to Neo4j Browser, or if you want to automate your data processing, it might be convenient to be able to use GDS procedures from Python. One possible approach is to use the Neo4j Python driver introduced in a preceding chapter (*Chapter 3, Characterizing a Graph Dataset*) and write code like this:

```
# driver instantiation
from neo4j import GraphDatabase
driver = GraphDatabase.driver(
    "bolt://localhost:7687",
    auth=("neo4j", "<PASSWORD>")
)

with driver.session() as s:
    # create projected graph named 'pG'
    s.run("CALL gds.graph.project('pG', 'NodeB', 'LINK')")
    # run Louvain algorithm on this projected graph
    # and stream results
    result = s.run("CALL gds.louvain.stream('pG')")
    # print results for all nodes
    print(result.data())
```

For GDS versions below 2.0, that was the only solution. But since GDS 2.0, the Neo4j team has released a Python client, allowing us, as data scientists, to call procedures without having to write any Cypher queries, just Python. Less error-prone, this method is definitely more elegant. One of the key features of this client is that it follows the same naming convention as the Cypher version. Let me show you what I mean by that.

In the following sub-section, we will install the GDS Python client and use GDS capabilities from Python: create projected graphs and run graph algorithms.

GDS Python principles

The code for this sub-section is available in the `Introducing_GDS_Python_Graph_Projection` notebook.

The GDS Python client is available in the `graphdatascience` package. You can install it with `pip`, from the command line:

```
pip install graphdatascience
```

Alternatively, you can install it from a Jupyter notebook:

```
!pip install graphdatascience
```

Then, we need to instantiate a `GraphDataScience` object. It needs to know how to access your database, so we must provide the same connection parameter we pass to a Neo4j driver object: the database URI and authentication credentials, which, in our case, is the username (using the default one, `neo4j`) and the password (the password that was set when creating the database in Neo4j Desktop):

```
from graphdatascience import GraphDataScience
gds = GraphDataScience(
     "bolt://localhost:7687",
     auth=("neo4j", "<PASSWORD>")
)
```

You can check the `gds` object has been properly instantiated and connected to your Neo4j database. For example, check the GDS plugin version is installed for your Neo4j instance with the following:

```
print(gds.version())
```

This is similar to what we have done with Cypher:

```
// Cypher
CALL gds.list()
```

We can also list all the available procedures from the GDS Python client with the following:

```
# Python
gds.list()
```

You might start to see a pattern here. Indeed, all procedures executed in Cypher with a query starting with `CALL gds.<something>...` are also available from the Python `GraphDataScience` object of the same name: `gds.something....`

Input and output types

Cypher procedures understand a few input types: strings, lists, maps, and numbers. Well, it's easy – it's the same for the Python client. For instance, consider the following Cypher query, using each of these types in different places:

```
// Cypher
CALL gds.graph.project(
     // a string
```

```
    "graphName",
    // a list
    ["Person", "Movie"],
    // a map
    {
        DIRECTED: {
            orientation: "UNDIRECTED",
            properties: [
                {
                    weight: {
                        defaultValue: 1.0
                    }
                }
            ]
        }
    }
)
```

In Python, it translates to the following:

```
# Python
gds.graph.project(
    # a string
    "graphName",
    # a list
    ["Person", "Movie"],
    # a dict
    {
        "DIRECTED": {
            "orientation": "UNDIRECTED",
            "properties": [
                {
                    "weight": {
                        "defaultValue": 1.0
                    }
                }
            ]
```

```
            }
        }
)
```

> **Warning**
>
> Notice that, while in Cypher, the quotes around the map keys raise errors – they are mandatory in Python's dictionary keys. That's why we can have `{LINK: ...}` in Cypher but have to use `{"LINK": ...}` in Python.

Both preceding code snippets create a projected graph filtering nodes by label (only nodes with `Person` or `Movie` will be included). It also filters relationships by the `DIRECTED` type, considering them to be undirected. It adds to the projected graph a relationship property whose name is `weight`, making sure all relationships in the projected graph will have a value for this property by setting a default value of `1`.

On the other end, the output format depends on the type of procedure, but it is also quite intuitive. As you saw with the `gds.list()` Python function, Cypher procedures returning multiple rows are returned as a `pandas.DataFrame` object in Python. A couple of other formats are also possible, consisting mainly of tuples of two values. We will discover them later.

Let's actually perform an analysis with the GDS Python client, step by step. Again, we will start by computing the degree distribution for our graph, by still taking advantage of the GDS but (almost) without writing a Cypher query. As with any GDS analysis, it all starts with creating a projected graph.

Creating a projected graph from Python

Creating a projected graph from Python is quite straightforward if you come from Cypher. Let's analyze the following example:

```
graph_object, graph_info = gds.graph.project(
    # projected graph's name
    "persons",
    # node projection
    "Person",
    # relationship projection
    {"KNOWS": {"orientation": "UNDIRECTED"}}
)
```

Here, we can note the following:

- On the first line, we are calling the `gds.graph.project` function. It is one of these functions with a special signature. Indeed, it returns a tuple of two elements:
  - The graph object: This is a Pythonic representation of the projected graph. It is important to keep it in a special variable, as we will need it later on.
  - Some information about the projected graph, which is similar to the ones you would see when executing the procedure with Cypher (see *Figure 6.1*).
- The function parameters are as follows:
  - A string containing the name of the projected graph.
  - The node projection configuration: This can be either a string (containing a node label), a list of strings (each of them being a node label), or a dictionary with a more detailed configuration. In this case, we are just using the `"Person"` string, which means the projected graph will only select nodes whose labels contain `Person`.
  - The relationship projection configuration: Here, again, this can be a single string (a relationship type), a list of strings (a list of relationship types), or a map. Here, we are selecting relationships of the `KNOWS` type, and telling the GDS that they should be considered undirected.

> **Warning about the projected graph name**
>
> The name of the projected graph has to be unique. If you try and create a new projected graph with a name already present in the graph catalog, you'll get an error saying that the graph already exists. You can check the content of the graph catalog using the `gds.graph.list()` function.

When creating a projected graph with Cypher, you should get the following result:

```
neo4j$ CALL gds.graph.project("persons", "Person", {KNOWS:
{orientation: "UNDIRECTED"}})
```

nodeProjection	relationshipProjection	graphName	nodeCount	relationshipCount	projectM
{ "Person": { "label": "Person", "properties": { } } }	{ "KNOWS": { "orientation": "UNDIRECTED", "aggregation": "DEFAULT", "type": "KNOWS", "properties": { }	"persons"	40948	715412	859

Started streaming 1 records after 38 ms and completed after 924 ms.

Figure 6.1 – The result of a graph projection procedure with Cypher

Running the preceding Python code will yield a result that is similar to the one displayed in the following screenshot:

	0
nodeProjection	{'Person': {'label': 'Person', 'properties': {}}}
relationshipProjection	{'KNOWS': {'orientation': 'UNDIRECTED', 'aggregation': 'DEFAULT', 'type': 'KNOWS', 'properties': {}}}
graphName	persons
nodeCount	40948
relationshipCount	715412
projectMillis	108

Figure 6.2 – Content of the graph_info variable

As you can see, it contains the same result as the one you would have obtained by running the procedure from Cypher: the graph configuration (that is, the name, node, and relationship projection as maps) and basic statistics (that is, the number of nodes and relationships).

But what if you need to use Cypher projections? Nothing too challenging here – just use the `gds.graph.project.cypher` Python function. The following code will create a projected graph with the same content as pgraph (which we created earlier), but using the Cypher projection instead of the native one:

```
graph_object, graph_info = gds.graph.project.cypher(
    # projected graph's name
    "personsCypher",
    # node projection
    "MATCH (n:Person) RETURN id(n) as id",
    # relationship projection
    "MATCH (a:Person)-[r:KNOWS]-(b:Person)
       RETURN id(a) as source, id(b) as target,
       COALESCE(r.weight, 1.0) as weight"
)
```

Native projection versus Cypher projection

When possible, prefer native projections over Cypher, as they are built to be much more effective in terms of the graph projection step. Cypher projections must be reserved for those cases where you have to create relationships or properties on the fly (that is, entities that do not exist in Neo4j but that you want to save in a projected graph). For instance, this might be the case when projecting a bipartite graph (for example, users and products with a *buy* relationship) into a monopartite graph of users, connected if and only if they bought the same product.

Once your projected graph is ready, it's time to run algorithms on it. In the next section, we will use the example of the GDS degree algorithm and use both the write and stream modes to deal with the computed metric.

Running GDS algorithms from Python and extracting data in a dataframe

In a preceding chapter, we learned that GDS algorithms offer multiple run modes, depending on where we want the results to be saved. In stream mode, the algorithm results are just streamed to the user, who has to decide what to do with them. In write mode, the results are persisted in the Neo4j database. Finally, mutate mode will update the in-memory projected graph with the results, which will be lost when the Neo4j instance is restarted, just like all the projected graphs. In this section, we will look at write and stream modes.

The code for the next paragraph is available in the Running_Algorithms_From_Python notebook.

write mode

As we just mentioned, when calling a GDS algorithm in `write` mode, the results of the algorithm computation will be written back to the main Neo4j graph. This is the only way to persist a result when the Neo4j server is restarted. The result can be either of the following:

- A node property: This is for all algorithms computing a metric for each node, such as centrality or community detection algorithms. We will see other examples in the next chapter.
- A new relationship: Some algorithms will add new relationships to the existing Neo4j database. For instance, this might be the case for algorithms computing similarity between nodes. For such algorithms, a relationship is added between the most similar nodes.

Let's go ahead and compute each node's degree using the GDS procedure, `gds.degree.write`. With Python, we have to use the following code:

```
# Python
gds.degree.write(
    graph_object,
    writeProperty="degree"
)
```

As you might have noticed, there are some differences between Cypher `CALL` statements and the GDS Python client syntax. Let's investigate them.

Key differences concerning the GDS with Cypher

From the preceding code snippet, you should notice some differences between the Cypher and Python syntax. As a reminder, the Cypher syntax for the same procedure is as follows:

```
// Cypher
CALL gds.degree.write(
    "persons",
    { writeProperty: "degree" }
)
```

There are two important differences to note, as follows:

- **Algorithm parameters**: These are passed as maps in Cypher, but they need to be provided as keyword arguments in Python.
- **Type of the projected graph parameter**: When using Cypher, we have to provide the graph name as a string, and the GDS takes care of finding the corresponding graph in the graph catalog. In Python, the functions expect a *graph object* as the first parameter.

This second bullet is the reason why we have saved the projected graph object returned by the `gds.graph.project` function in the first place. But if you have lost the graph object, you can still retrieve it if you remember its name, using the Python-specific function:

```
graph_object = gds.graph.get("persons")
```

> **Note**
>
> In the notebooks linked to this book, I am using a custom function called `created_projected_graph` to be sure the projected graph does not already exist (which raises exceptions).

Going back to the degree function, we can see that it returns the following results, identical to the ones returned by the Cypher procedure:

```
nodePropertiesWritten                                                   40948
centralityDistribution    {'p99': 111.00048065185547, 'min': 0.0, 'max':...
writeMillis                                                               348
postProcessingMillis                                                      165
preProcessingMillis                                                         0
computeMillis                                                               0
configuration             {'jobId': '0a87bd0e-0de6-4a72-a891-7341b70e55e...
Name: 0, dtype: object
```

Figure 6.3 – The result of the degree algorithm in write mode

It does not show the degree for each node, but only some statistics about the algorithm: the runtime, the number of properties written to Neo4j, and some percentiles of the computed metric (the degree).

To access the computed degree for each node, we have to retrieve them from Neo4j, using Cypher. To do so, we can use another useful function from the GDS Python client: `gds.run_cypher`. This lets us run a Cypher query without having to instantiate a Neo4j driver on our own since the GDS Python package already holds an instance of such a driver. Let's fetch the `id` and `degree` properties for all nodes with the `NodeB` label using the following code:

```
degrees = gds.run_cypher("""
MATCH (n:Person)
RETURN id(n) as node_id, n.name as name, n.degree as degree
""")

degrees.head(10)
```

Note that `degrees` is a pandas dataframe, where each row contains the `id` property and computed degree for each node. The following figure shows the first five rows of this dataframe:

	node_id	name	degree
0	9046	Kirsten Johnson	0.0
1	9047	Julien Leclercq	24.0
2	9048	Mike Flanagan	33.0
3	9049	Robert Cullen	11.0
4	9050	José Luis Ucha	11.0

Figure 6.4 – The first five rows of the degrees dataframe

Before using any results, let's investigate how to retrieve algorithm results in streaming mode.

stream mode

When streaming results, the computed values are directly accessible and can be stored inside a variable:

```
degrees = gds.degree.stream(
    graph_object,
)
```

An overview of the new degrees dataframe is shown in the following screenshot:

	nodeId	score
0	9046	0.0
1	9047	24.0
2	9048	33.0
3	9049	11.0
4	9050	11.0

Figure 6.5 – The result of the degree algorithm in stream mode

But an important difference here is that the `nodeId` column is not the same as the `id` column from our previous dataframe, which was obtained by fetching the results after a `write` mode. Indeed, this `nodeId` column is the internal `nodeId` value of the GDS projected graph. It has nothing to do with the Neo4j internal `id` property that we obtain with the `id()` function. The `nodeId` value returned by the GDS should not be used anywhere outside of the GDS. In order to map this `nodeId` value to the persisted node object in Neo4j, we would use the `gds.utils.asNode(nodeId)` function in Cypher, and then access its properties like this:

```
CALL gds.degree.stream("persons") YIELD nodeId, score
WITH score as degree, gds.util.asNode(nodeId) as node
RETURN id(node) as id, degree
```

In Python, we can do something similar. Let's take a look at the code for one row:

1. Extract one row of the `degrees` dataframe. Here, we have selected the second row:

   ```
   row = degrees.iloc[1]
   ```

2. Call the `gds.utils.asNode` function to retrieve a node object:

   ```
   node = gds.util.asNode(row["nodeId"])
   ```

   Here, node is of the `neo4j.graph.Node` type, which is a type defined in the Neo4j Python driver to represent a node as a Python object.

3. Get the Neo4j's `id` property of this node:

   ```
   node.id
   ```

   We could repeat this operation for each node, but that would result in a database hit for each node, which is quite inefficient. Fortunately, the GDS also exposes another function, `gds.utils.asNodes`, by accepting a list of `nodeId` values as a parameter and returning a list of node objects. We can then update our dataframe in two steps, as defined next.

4. First, fetch the node objects for all values in the `nodeId` column at once:

   ```
   nodes = gds.util.asNodes(degrees["nodeId"].to_list())
   ```

5. Then, extract the `id` property for each of these node objects:

   ```
   degrees["id"] = [node.id for node in nodes]
   ```

   This time, the `id` column in our `degrees` dataframe (see the following screenshot) matches the `id` property of the node. It can be used to match other node properties, if any, or any external systems that this `id` property might be connected to:

	nodeId	score	id
0	9046	0.0	9046
1	9047	24.0	9047
2	9048	33.0	9048
3	9049	11.0	9049
4	9050	11.0	9050

Figure 6.6 – The degrees dataframe after translating the id property

Why not always use write mode?

You might ask yourself since using the algorithm in `stream` mode requires a post-treatment to be usable, why might you want to use it? There are several reasons for this. One of them is that you do not need the real node data, for instance, for data visualization. Another reason is that the user you use to establish a connection to Neo4j does not have write permission for the database – in that case, you will have to stream the results.

Dropping the projected graph

Once you are done with the projected graph, meaning you have run all the relevant algorithms for your analysis and have saved the results elsewhere, don't forget to drop the projected graph. Since it is all saved in memory, having too many projected graphs that are no longer useful might cause some damage that could easily be avoided. Applying this good habit from the beginning will save you some trouble. Dropping the projected graph with Python is achieved with the following line:

```
gds.graph.drop(graph_object)
```

Also, be aware that, when using large graphs, you can estimate the size of the projected graph and check whether your machine runs with enough RAM with the following functions/procedures:

```
gds.graph.project.estimate(<nodeProjectionConfig>, ...)
gds.graph.project.cypher.estimate(<nodeQuery>, ...)
```

Now that we know how to run a GDS algorithm from Python and retrieve it in a dataframe, building an ML model from it should not be a problem. Let me give you an example in the following section, where we will illustrate how we can use graph-based features in a model training phase.

Using features from graph algorithms in a scikit-learn pipeline

Now we have all the necessary knowledge to actually use graphs for ML. In this section, we are going to wrap everything up using the GDS Python client to create features and extract data into a dataframe that can be fed into a `scikit-learn` model training pipeline.

But before we get to this, let me give you an overview of the ML possibilities with graphs.

Machine learning tasks with graphs

In general, ML comprises several types of tasks on various kinds of objects: from sales predictions with time series analysis to patient diagnosis thanks to medical imagery to text translation in many languages with **natural language processing** (**NLP**), ML has proven its usefulness in many situations.

In each of these cases, you have to build a dataset made of **observations** (usually, the rows). Each observation has a certain number of characteristics or **features** (that is, the columns of your dataset). Depending on the task, you can also have a **target**: the value (in the case of a regression) or class(es) (in the case of classification). The target is the variable your model is trying to predict.

When your observations are stored as nodes in a graph, the process is exactly the same: you extract features from the node's properties to build your dataframe. The sole difference is that the graph structure can provide more features: the results of the graph algorithms we studied in earlier chapters give us some information about the observation's *role* or *position* in the graph with respect to the other nodes. Similarly to usual tabular datasets, both regressions and classifications can be performed in this way.

But graphs do not only contain nodes. They are also made of edges or links. And in a time-evolving graph, both nodes and edges are affected. An important graph-specific task in ML is **link prediction**. *Chapter 9, Predicting Future Edges*, is dedicated to this particular topic.

In this section and the next two chapters, we will focus on the first kind of ML task: node classification.

The code for this section can be found in the `Training_SK_Model` notebook of the `Chapter06` folder, which is in the GitHub repository for this book.

Time for some practice!

Our task

We will be working with our graph made of nodes with the `MainComponent` label, a subset of `Person` nodes belonging to the main component. In order to make it a bit more interesting, we will start by adding some properties to our nodes.

Note about the results in this section

The data we are studying here was generated only for this chapter. It is definitely biased, and no conclusion should be drawn from the prediction results. However, there are plenty of examples showing how the use of such features, output by graph algorithms, can, sometimes considerably, increase a model's predictive power. Some of these examples are given in the *Further reading* section.

Let's use our graph knowledge to compute some additional features.

The code for this last section can be found in the `Training_SK_Model` notebook.

Computing features

As we did in the preceding sections, we are going to create a projected graph and compute the Louvain communities in the graph (see *Chapter 4, Using Graph Algorithms to Characterize a Graph Dataset*, for an introduction to the Louvain algorithm). I'll just repeat the code here. For detailed explanations, refer to the preceding sections:

```
from graphdatascience import GraphDataScience
from utils import create_projected_graph

gds = GraphDataScience(
     "bolt://localhost:7687",
     auth=("neo4j", "<PASSWORD>")
)
graph_object = create_projected_graph(
     gds,
     graph_name="persons",
     node_spec="Person",
     relationship_spec={"KNOWS": {"orientation": "UNDIRECTED"}}
)
gds.louvain.write(
     graph_object,
     writeProperty="louvain"
)

gds.graph.drop(projected_graph_object)
```

After computing the Louvain communities for our graph and storing the community each node belongs to in the graph, we can extract the data and build a model.

Extracting and visualizing data

To extract the data, we will use the gds object that exposed a run_cypher method:

```
data = gds.run_cypher(
    """MATCH (person:Person:MainComponent)
    -[:IS_CITIZEN_OF]→(country:Country)
    WITH   person, country
    MATCH    (person)-[:ACTED_IN|DIRECTED]->(movie:Movie)
    WITH   person, country,
    count(DISTINCT movie) as numberOfMovies
    RETURN id(person) as id,
    country.name = 'United States of America' as target,
    numberOfMovies as nbMovies,
    person.louvain as louvain
    """
)
data.head()
```

The data variable contains a dataframe with four columns: id, louvain, nbMovies, and target.

We can check the content of the target column and observe a clear **imbalance** in our dataset, as it contains more False observations than True ones:

```
False     18587
True       7609
Name: target, dtype: int64
```

We will have to take this into account when building our classification model.

Building the model

As we observed in the preceding section, the dataset is slightly imbalanced. Before starting to train a model, let's fix this by down-sampling the false observations.

Down-sampling

scikit-learn provides the tools to perform such a down-sampling. It is performed in the following code. First, true and false observations are split, then the false observations are down-sampled before being merged again with the true observation to create a fully balanced dataset we can use in the rest of the analysis:

```
from sklearn.utils import resample

true_data = data[data.target == 1]
false_data = data[data.target == 0]

false_data_downsampled = resample(
     false_data,
     replace=True,
     n_samples=len(true_data),
     random_state=1234,
)
data = pd.concat(
     [true_data, false_data_downsampled],
     axis=0
)
```

Our data is now ready to perform the analysis.

Training a classification model

In this chapter, we will use a simple `DecisionTreeClassifier`. We will drop the hyperparameters tuning phase and just split our dataset into a train sample and a test sample:

```
from sklearn.model_selection import train_test_split
X = data[["nbMovies", "louvain"]]
y = data.target

X_train, X_test, y_train, y_test = train_test_split(
     X, y, test_size=0.33,
     stratify=data.target,
     random_state=1234
)
```

Since we are going to compare several sets of features (using `nbMovies` or `louvain` separately, or both together), we create a small helper function that will train a classifier on the train sample and compute and return its score on the test sample:

```
from sklearn import tree
def train_and_score(
    X_train, y_train, X_test, y_test,
    feature_columns
):
    X_train_bis = X_train[feature_columns]
    X_test_bis = X_test[feature_columns]
    clf = tree.DecisionTreeClassifier()
    clf = clf.fit(X_train_bis, y_train)
    return clf.score(X_test_bis, y_test)
```

Thanks to this function, we can test our model with the following `for` loop:

```
feature_sets = [
   ["nbMovies"],
   ["nbMovies", "louvain"],
]

for feature_set in feature_sets:
    score = train_and_score(
        X_train, y_train, X_test, y_test,
        feature_set
    )
    print(f"{str(feature_set):30} => {score:2.2f}")
```

The printed results are as follows:

```
['nbMovies']                         => 0.544
['nbMovies', 'louvain']              => 0.787
```

As you can see, in this particular example, the Louvain partitioning adds some information to our model and helps us increase the score of our classifier. This is something quite remarkable: the Louvain algorithm only knows about the person's relationships with each other, for instance, who has worked with whom. This information alone is sufficient to get almost 80% accuracy in a classification task, without any other knowledge.

Summary

In this chapter, you learned about the GDS Python client. From graph management (projection, retrieval, and deletion), to running algorithms and retrieving their results in a pandas dataframe, and all we have done in the preceding chapters with Cypher, you are now able to do it without needing to open the Neo4j browser anymore. By only using a Jupyter notebook, you can take advantage of the full power of Neo4j and the GDS. Since the GDS procedures return pandas dataframes, it is quite straightforward to include these results within a Python ML pipeline, for instance, by using `scikit-learn`, as we have done in the last section of this chapter.

This chapter and the preceding ones have shown you how to extract features from a graph dataset, taking advantage of the graph structure. Features such as a degree, or more generally, centrality metrics, and community ID are only available if you consider the relationships between the entities in your dataset to build a graph. Depending on the problem you are trying to solve, these relationships can considerably increase your model performance.

In the next chapter, we are going to dive into automatic feature extraction for graphs, that is, graph embeddings. Similarly to texts or images, graphs are high-dimensional datasets. Reducing their dimensionality, while preserving the information they convey is challenging, and different techniques exist to solve this challenge. The GDS contains implementation for the most popular of these techniques, and we will learn how to take advantage of them by continuing to use the GDS Python client.

Further reading

If you want to know more about the topics covered in this chapter, here are some recommended readings:

- *Neo4j GDS Python client manual*: `https://neo4j.com/docs/graph-data-science-client/current/`
- *Graph Algorithms*, by M. Needham, A. E. Hodler (O'Reilly)

Exercise

In order to get more proficient with the GDS Python client, I recommend you redo the analysis from *Chapter 4, Using Graph Algorithms to Characterize a Graph Dataset*, with it. That is, do the following:

1. Build projected graphs that contain the following:

   A. `Person` nodes and all relationships.

   B. `Person` and `Movie` nodes with all relationships.

   C. `Person` nodes with `KNOWS` relationships, undirected.

   D. `Person` nodes with `KNOWS` relationships aggregated to keep one single relationship between the same two nodes.

E. `Person` nodes with `KNOWS` relationships aggregated to store the number of relationships between the same two nodes in a `weight` property.

F. Using Cypher projection: `Movie` nodes and a relationship between two movies when at least one person acted in or directed both movies. Save the movie's release year and the relationship's weight (the number of persons collaborating on both movies) as node and relationship properties, respectively.

2. Run graph algorithms: Using the projected graph with `Person` nodes and `KNOWS` relationships, compute the following:

   A. Node's degree using `stream` mode. Study the influence of the `orientation` parameter (*Chapter 4, Using Graph Algorithms to Characterize a Graph Dataset*, in the *Algorithm configuration* section).

   B. Run the **weakly connected component** (**WCC**) algorithm in write mode. Within Python, retrieve the largest community ID.

7
Automatically Extracting Features with Graph Embeddings for Machine Learning

When dealing with a graph dataset, we can rely on feature engineering and define important features for our context, taking into consideration the graph structure via features extracted from graph algorithms such as node importance or belonging to a community. However, as for other kinds of complex objects—images or texts, for instance—there are ways to automatically extract features from a graph. They are called graph embedding algorithms, and they are able to retain part of the graph structure while representing objects in a low-dimensional space. In this chapter, we will introduce several of these algorithms, which can be used from the Neo4j **Graph Data Science** (**GDS**) library: Node2Vec and GraphSAGE. On one hand, Node2Vec is inspired by the Word2Vec text embedding algorithm and only works when the full dataset is known beforehand, meaning we won't be able to predict embeddings of new nodes when the graph grows with time. On the other hand, GraphSAGE uses a different approach and is able to compute embeddings for unseen nodes. This will be the opportunity for us to introduce another feature of GDS: the model catalog, which can store trained models so that they can be reused later on.

In this chapter, we're going to cover the following main topics:

- Introducing graph embedding algorithms
- Using a transductive graph embedding algorithms
- Training an inductive embedding algorithm
- Computing new node representations

Technical requirements

In order to be able to reproduce the examples given in this chapter, you'll need the following tools:

- Neo4j 5.x installed on your computer (see the installation instructions in the first chapter)
- **Graph Data Science** plugin (version >= 2.2)
- A Python environment with Jupyter to run notebooks
- An internet connection to download the plugins and the datasets
- Any code listed in the book will be available in the associated GitHub repository (`https://github.com/PacktPublishing/Graph-Data-Science-with-Neo4j`) in the corresponding chapter folder

> **Code examples**
>
> *Unless otherwise indicated, all code snippets in this chapter and the following ones use the GDS Python client.* Library import and client initialization are omitted in this chapter for brevity, but a detailed explanation can be found in the *Introducing the GDS Python client* section of *Chapter 6, Building a Machine Learning Model with Graph Features*. Also, note that the code in the code bundle provided with the book is fully runnable and contains all imports.

Introducing graph embedding algorithms

This section will introduce the principles of graph embeddings and explain the idea behind two of the most famous algorithms: **Node2Vec** and **GraphSAGE**. In the following sections, we will use the GDS implementation of these two algorithms to extract embeddings for nodes stored in a Neo4j database.

Defining embeddings

Machine learning (**ML**) algorithms—classification or regression—require an input matrix made of observations (rows) and features (columns). While this is trivial for tabular datasets (for example, the Iris or Titanic datasets), this is already a challenge for datasets made of more complex objects such as texts, images, or graphs. The question is: how can we build a matrix from these objects while preserving their nature? By nature, we mean here the key characteristics that will not kill the predictive power of our data. In the case of texts, this is the meaning of the sentences or words. We will see later in this chapter that the nature of a graph does not have a unique definition either.

To start with, let's understand which types of objects we want to embed in a graph dataset.

Graph embedding vocabulary

Complex objects can be represented as high (sometimes very high)-dimensional matrices. For instance, texts can be vectorized thanks to a vocabulary matrix of size *N* where *N* is the number of different words in the whole dataset, which you can imagine can be huge (several tens of thousands or even more).

In the case of graphs, several objects can actually be embedded:

- **Nodes**: This is the most common case and the one we are going to study in the rest of this chapter. Nodes are our observations, and we are trying to capture somehow their role or position in the graph with embedding.
- **Graph**: In some circumstances, the objects of interest are not the individual nodes but the whole graphs. Think about chemistry where molecules are represented by a graph. Researchers are interested in finding molecules with similar structures, regardless of the atoms making them up.
- **Edge**: This is the last kind of object we can find in a graph, and it can also be represented by a vector in situations where edges are the observations. This happens in link prediction problems, for instance. We will detail this type of graph challenge in a later chapter, *Chapter 9, Predicting Future Edges*.

In this book, we will only talk about *node embedding*, referred to simply as embedding, as in many other writings on the topic.

Embedding techniques

Going back to our initial problem, how can we then create a matrix with *N* rows (the number of nodes) and where each column is a feature? The first approach is to use the **adjacency matrix** of the graph. Quickly introduced in *Chapter 1, Introducing and Installing Neo4j*, an adjacency matrix is an *NxN* matrix, where *N* is the number of nodes in the graph. The *ij* element of the matrix contains the weight of the edge between nodes *i* and *j*. Or, with an equation, the adjacency matrix *A* of a weighted graph with a set of vertices denoted *V* is as follows:

$$A_{ij} = w_{ij} \forall (i,j) \in VxV$$

For instance, considering the graph represented on the left-hand side of the following diagram, we would obtain the adjacency matrix represented on the right-hand side, where nodes are sorted in alphabetical order:

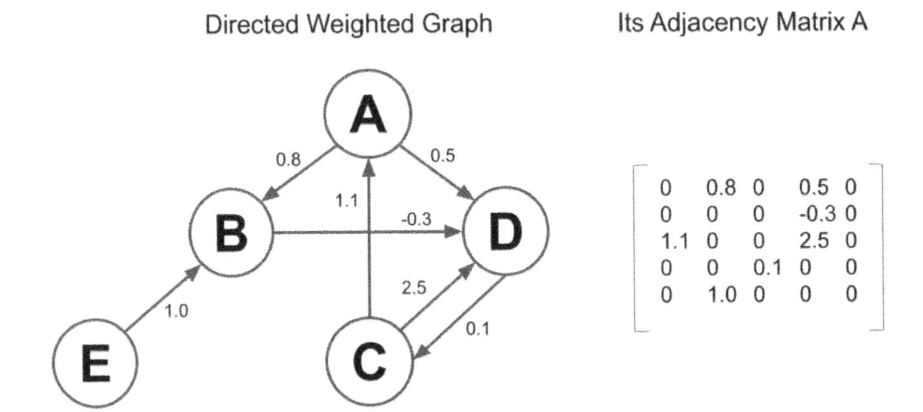

Figure 7.1 – Example of a graph and its adjacency matrix

Using the adjacency matrix as input to train an ML model has several caveats:

1. By definition, the adjacency matrix of a graph with *N* nodes contains N^2 items. For sparse graphs, we can imagine using sparse matrix representation, but this is not always possible. And we know that dealing with matrices with millions of elements, or more, is not manageable by current systems: computations will either collapse due to **out-of-memory** (**OOM**) errors or take days to converge, in the best-case scenario.
2. But let's imagine we can overcome this challenge, by taking advantage of parallelism, using better hardware, or simply smarter implementation. What can't be overcome, however, is the fixed number of features introduced by the adjacency matrix: remember that in an ML model, rows are observations and columns are features. Using the adjacency matrix as input, we would have *N* features, *N* being the number of nodes. Now, what if we add a new node to our graph? The adjacency matrix grows to reach size *(N+1)x(N+1)*, adding one observation but, most importantly, adding one feature. This means that the model we trained with *N* features is not usable anymore! While this is manageable in some circumstances where you know the whole graph beforehand and do not need to make predictions on unseen nodes, many use cases won't be covered by this technique.

These are some of the reasons why embedding techniques have been developed. Their goal is to reduce the size of the vector representation of the objects, to overcome problem number *1*, ideally in an inductive way, making them able to make predictions on unseen nodes.

The challenge when reducing dimension is to do so *without losing important information* and preserve the *key characteristics* of the objects of interest. *Key characteristics* is a term that is a bit vague on purpose: it depends on the final goal of the analysis.

Consider the following screenshot, which you may have already seen in a different context:

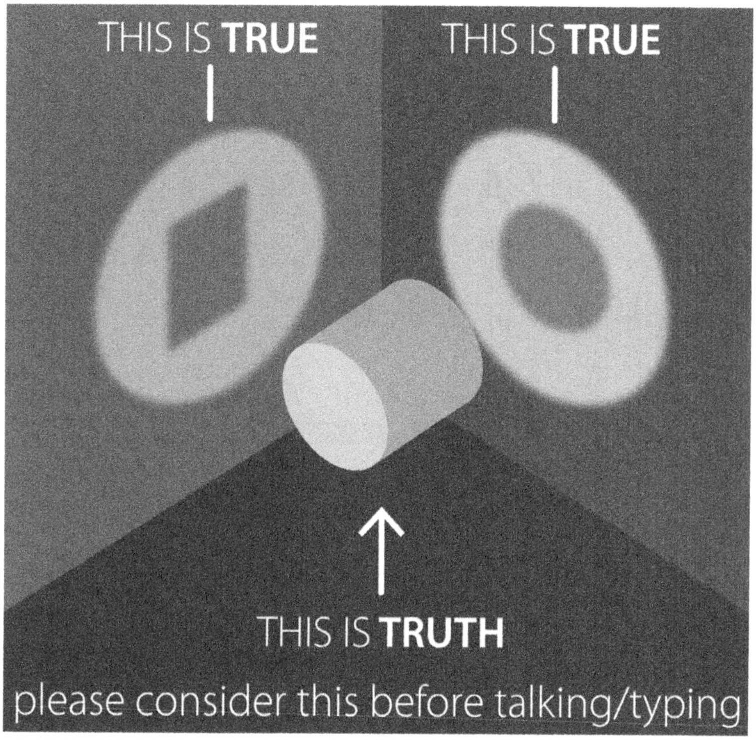

Figure 7.2 – Describing a cylinder with its projections

The real object is the 3D cylinder shown at the center of the image. Both shadows—the square on the left and the circle on the right—show a part of the information of the cylinder on a 2D plane. Each contains partial information about the full object, but it can be enough depending on our assigned task.

Let's imagine the cylinder is a can, filled with soda, soup, or your preferred meal. If the question is: "Can I store this can vertically in my fridge?", then you only need to know the cylinder's height, and the square projection contains all the information you need to answer this question. On the other hand, if the question is "How many cans can fit vertically on this pallet?", then you'll probably need the information contained in the circle since the cylinder diameter is crucial.

This is what embedding is all about: projecting a dataset (the cylinder) into a lower-dimension space (the circle or the square) while preserving some aspects of the initial dataset (its diameter or height).

In any case, the lower the embedding dimension, the lower the preserved information about the original object. Embedding is always a trade-off between the information we can afford to discard and the size of the matrix we are using so that computation will complete on your hardware in a reasonable time scale.

Going back to our node embedding problem, the goal of such an algorithm will be to assign a low-dimensional vector. The usual order of magnitude is a few hundred or fewer dimensions to each node (see the following diagram):

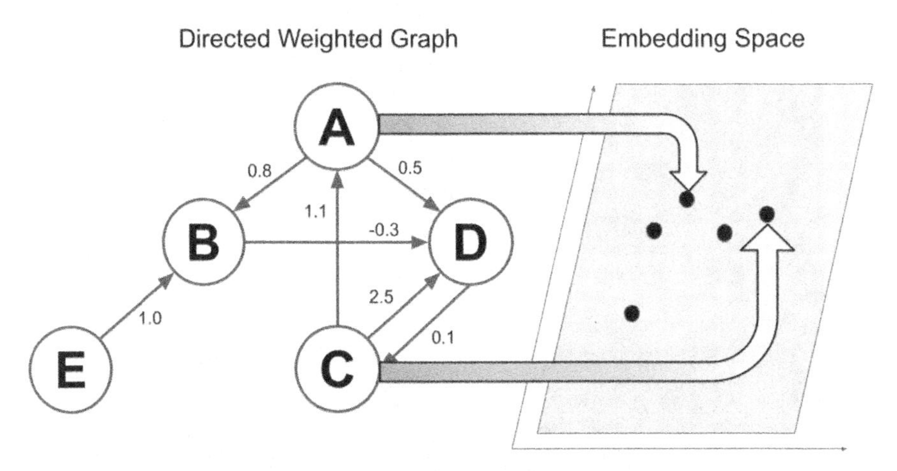

Figure 7.3 – Embedding principles

But this has to be done in a way that some information about the graph is still contained within this representation. We will see in the next section what kind of information can be preserved.

Graph embedding classification

There are many different ways to reduce a matrix dimension. As in many situations, there are also many ways to classify them. The following subsections introduce some embedding algorithm properties that can be used to characterize and classify them.

Transductive or inductive

We already talked about inductive algorithms earlier in this chapter but they are truly important, so let me recall and clarify the definitions for inductive algorithms and their counterpart, transductive algorithms:

- **Transductive**: A transductive algorithm is one that needs all the dataset as input, meaning it won't be able to make predictions on new observations (nodes) without recomputing everything
- **Inductive**: On the opposite side, this algorithm is able to make predictions on new observations based on its learning from a training dataset

Among the algorithms we are going to study, Node2Vec is transductive, while GraphSAGE is inductive.

Positional or structural

An important question to ask ourselves when dealing with graph embedding is: what kind of information do I want to encode? As for our cylinder, node embedding can capture different kinds of information:

- **Positional**: Encode node position in the graph, such that two nodes close to each other in the original graph have close vector representation in the embedding space
- **Structural**: Encore node neighborhood, such that two nodes with similar vicinity have similar embeddings, even if the two nodes are *far* from each other in the original graph

To understand the difference, look at the following diagram:

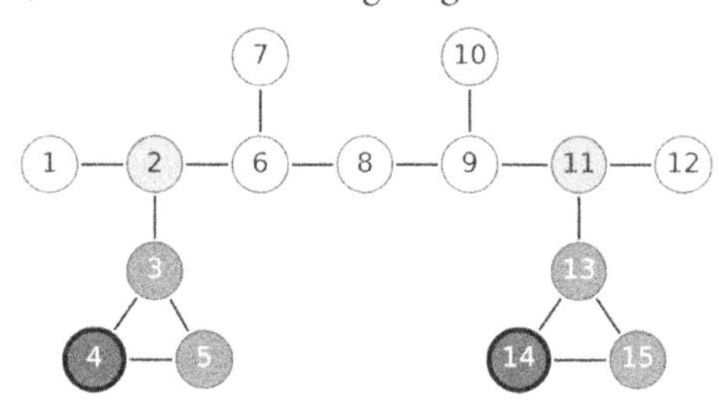

Figure 7.4 – Example graph: a lighter shade means the node is further away from node 4 (or 14)

The question is: should the embedding of node `4` be closer to the one of node `3` or the vector representation of node `14`? While `4` and `3` are very close to each other in the graph, the neighborhood of `4` is much more similar to the one of `14` due to the graph symmetry (topology).

The following screenshot is depicting one possible generated embedding vector for this graph. You can see that this embedding algorithm puts nodes `4` and `14` very close to each other, while node `3`, close to node `13`, is further apart in the embedding space:

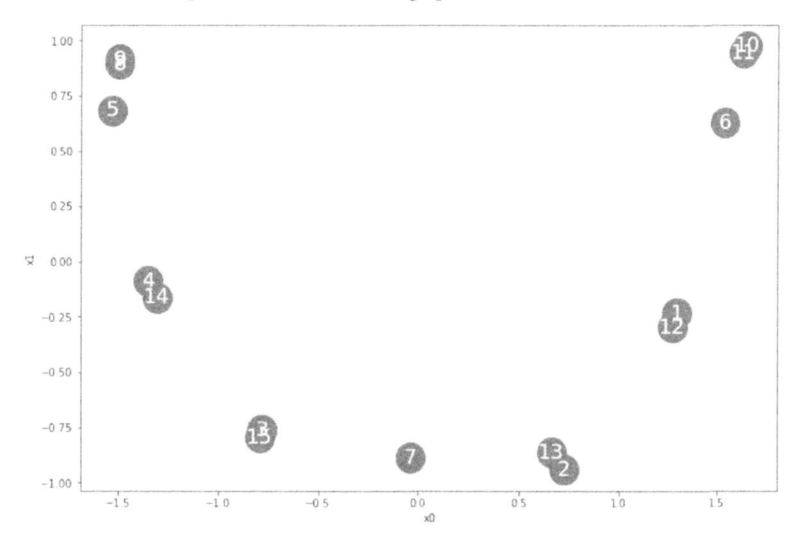

Figure 7.5 – Example structural node embeddings

This algorithm has hence encoded the structural nature of the graph.

Another algorithm generates the embeddings represented in the following screenshot:

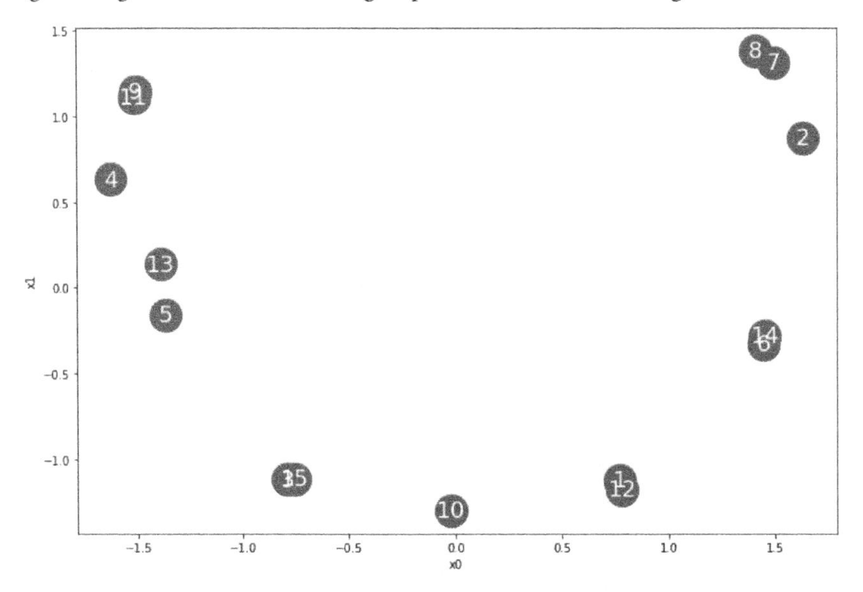

Figure 7.6 – Example positional node embeddings

Contrarily to the first example, here we can see that nodes `4` and `14` are very far from each other: this algorithm tends to encode the positional information, and nodes with similar neighborhoods tend to be closer in the embedding space.

Generally speaking, embedding techniques based on the adjacency matrix such as matrix factorization or random walks (for example, Node2Vec) are known to belong to positional embeddings. That's due to the fact that they rely on walks, hence proximity information. On the other hand, graph neural networks, using message-passing techniques, tend to capture the structural embeddings. We can, however, note that research is ongoing to try to create algorithms that would be able to merge both approaches (see, for instance, the **position-aware graph neural network** (**P-GNN**) algorithm from the **Stanford Network Analysis Project** (**SNAP**) group, referenced in the *Further reading* section at the end of this chapter).

Finally, be aware that some embedding algorithms are able to take into account node properties together with the graph structure, but this is not always the case. For the algorithms we are going to talk about in the rest of this chapter, only GraphSAGE has such an ability.

That being said, let's start to understand how graph embedding algorithms actually work. We'll start with the transductive embeddings and the Node2Vec algorithm.

Using a transductive graph embedding algorithm

As we stated in the preceding section, a transductive algorithm is characterized by the fact that it works only on a full dataset, meaning it won't be able to make any predictions on new observations. But, as with the centrality or community detection algorithms we have already crossed in the preceding chapters, these algorithms can be useful in circumstances where your graph is not evolving too fast. The GDS library currently contains two such algorithms: **Node2Vec** and **Fast Random Projection** (**FastRP**). We'll describe the principles and usage of the Node2Vec algorithm. The usage of the FastRP algorithm will be very similar.

Understanding the Node2Vec algorithm

The Node2Vec algorithm is derived from the **DeepWalk** algorithm. In order to understand DeepWalk, we also need to know about the **Word2Vec** and **SkipGram** models.

As you can imagine, Word2Vec is an embedding algorithm for words within texts. As for a graph, a text can be represented by a high-dimension matrix encoding each different word in the text—the vocabulary. But this matrix is way too large to be analyzed. To reduce the dimension of the word vector representation, the Word2Vec algorithm is used. Its steps can be summarized as follows:

1. For each word, create tuples based on neighbor words, parametrized by the size of the window, `w`. For instance, given the input text `Rome was not built in a day` and a `w=5` window size, we would generate the following pairs (where the window is centered on the word of interest):

   ```
   Rome was
   Rome not
   was Rome
   was not
   was built
   ...
   a built
   a in
   a day
   day in
   day a
   ```

2. A neural network is trained with the objective of *predicting the probability for a target word to be a neighbor (within the window size) of an input word* using the preceding `(input, target)` pairs. This neural network uses the so-called **SkipGram** architecture with a single hidden layer of dimension *d*, corresponding to the embedding size. If you want to learn more about SkipGram, some references are given in the *Further readings* section at the end of this chapter.

The DeepWalk algorithm, presented in 2014 by Perozzi et al., uses the same approach, but for graphs. To make the parallelism with texts, *sentences* are created by random walks in the graph: from a starting node, we can traverse relationships to the node neighbors up to l hops. Once the sentences of nodes are generated, we can create pairs of adjacent nodes in a window of size w and pass these pairs to the skip-gram model.

DeepWalk is parametrized by the following:

- The length of the walk, l
- The window size, w
- The number of walks generated from a given node, k: the higher k, the slower the training

Additionally, we need to define some parameters for the neural network (skip-gram) training:

- The embedding size, equal to the dimension of the hidden layer in the skip-gram architecture

This is how DeepWalk is built, but what is the relation with Node2Vec? Node2Vec is a variation of DeepWalk, in the way the random walks are performed. It introduces two new parameters:

- *Return parameter p*: Controls the probability of returning to the previous node. This probability is proportional to 1/q, such that high values of p reduce the probabilities to return to the node that's just been visited, extending the *radius* of each random walk: when p is low, the walk tends to stay close to the start node since back and forth are favored.
- *In-out parameter q*: Its name comes from its ability to control the choice between inward (somehow local) and outward nodes. With high values of q, the algorithm will tend to choose nodes close to the previous node in the walk, somehow mimicking the breadth-first traversal of a graph.

Consider the following diagram:

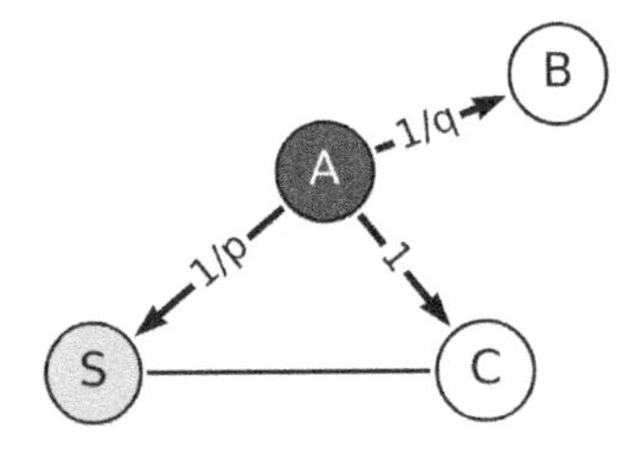

Figure 7.7 – Node2Vec return and in-out parameters illustration

The walk, started from node **S**, is now on node **A**, and the algorithm has to choose which node goes next between **B**, **C**, and going back to **S**. The numbers on the relationships indicate the probabilities used by Node2Vec to generate the path:

- *1/p* to return to the previously visited node (**S**)

- For other nodes:
  - It uses a probability of 1 if the length of the shortest path between **S** and the node is 1 (**C**)
  - The probability is set to *1/q* otherwise

In this way, the random walks can be biased toward breadth-first exploration ($q >> 1$) or toward a more local exploration ($p << 1$).

Enough of theory—let's see how we can actually compute node embeddings using the GDS library in the next section.

Using Node2Vec with GDS

Similarly to the other algorithms we've used in preceding chapters (for example, the Louvain algorithm for community detection), embedding algorithms run on a projected graph and have several run modes: `stream` to directly extract results without persistence, `write` to persist results as node properties in Neo4j, or `mutate` to alter the projected graph. In this section, we are going to use the `stream` mode only. In order to understand what's going on and visualize it better, we are going to leave our Netflix dataset to one side for a while and use a toy example instead.

The data we are going to use is stored in the `data/sample_graph.edgelist` file in the code bundle coming along with the book. Each row contains two numbers: the IDs of nodes that need to be connected. Let's import it:

1. First, create a new database called **Sample Graph** in Neo4j Desktop (see *Chapter 1, Introducing and Installing Neo4j*, for instructions).
2. Install APOC and the GDS plugins (see *Chapter 2, Using Existing Data to Build a Knowledge Graph*, for details about how to install APOC; the same procedure can be used for GDS).
3. Copy the data file into the `import` folder of this brand new database (also see *Chapter 2, Using Existing Data to Build a Knowledge Graph*, if you need more how-to).
4. Import the data with the following query:

```
LOAD CSV FROM "file:///sample_graph.edgelist" AS row
FIELDTERMINATOR ' '
MERGE (a:Node {id: toInteger(row[0])})
MERGE (b:Node {id: toInteger(row[1])})
MERGE (a)-[:RELATED_TO]-(b)
```

This sample graph contains 789 nodes with a `Node` label with `20717` unweighted relationships of type `RELATED_TO` that can be considered undirected. Let's see what we can say about it.

We are going to compute the Node2Vec embeddings for each node of the graph and try to visualize the result. The code for this section is available in the `Node2Vec` notebook on GitHub, in the folder associated with this chapter.

As with any other algorithm, embeddings are run on the projected graph, so let's start by creating one:

1. Import the `create_projected_graph` helper function, which will drop any existing projected graph with the same `graph_name` value if it already exists to prevent errors and create a new projected graph:

```
from utils import create_projected_graph
```

2. Create a projected graph with the name `pgraph`, containing all nodes with a `Node` label and all relationships with type `RELATED_TO`, considering them as undirected:

```
graph_object = create_projected_graph(
        gds,
        graph_name="pgraph",
        node_spec="Node",
        relationship_spec={"RELATED_TO": {"orientation":
"UNDIRECTED"}}
)
```

3. Run the Node2Vec algorithm on this projected graph:

```
res = gds.beta.node2vec.stream(
        graph_object,
        walkLength=20,
        embeddingDimension=18,
        inOutFactor=10,
)
res.head()
```

The returned `res` variable is a DataFrame containing two columns: `nodeId` (the GDS internal node ID) and `embedding`, which itself is a list of 18 elements, controlled by the embedding dimension. The first two rows are reproduced here:

	nodeId	embedding
0	0	[-1.321009874343872, -0.17161105573177338, -0.03883824124932289, -0.09094082564115524, 0.22463375329971313, 0.018603039905428886, 0.0418054535984993, -0.2145107537508011, 0.03011900559067726, -0.185122549533844, 0.10252618789672852, 0.04060518369078636, 0.0441429428756237, 0.07056885957717896, 0.21127858757972717, 0.17849786579608917, -0.018250489607453346, -0.14861491322517395]
1	1	[-1.427160382270813, -0.21028603613376617, 0.0920993983745575, -0.09148993343114853, 0.40076154470443726, 0.07538198679685593, 0.02744310162961483, -0.21002282202243805, -0.04245203360915184, -0.24959523975849152, 0.11915363371372223, 0.12276660650968552, 0.10642865300178528, 0.29195764660835266, 0.4075036942958832, 0.41647404432296753, 0.2468593716621399, -0.23009197413921356]

Figure 7.8 – First two rows of the DataFrame returned by the gds.beta.node2vec.stream function

Let's analyze this result a bit further. First, we'll explode the embedding list into columns so that we end up with one column per feature as usual:

1. First, we create a list of `list` from the `res.embedding` series:

```
embedding_list = res.embedding.to_list()
```

2. Then, we create a list of column names that will be `e_0, e_1, e_2...`:

```
f_names = [f"e_{k}" fork in range(len(embedding_
list[0]))]
```

3. Finally, we can append the columns to the `res` DataFrame:

```
res.loc[:,f_names] = embedding_list
```

The first row of the `res` DataFrame now looks like this:

	nodeId	embedding	id	e_0	e_1	e_2	e_3	e_4	e_5	e_6	...	e_8	e_9	e_10	e_11	
0	0	[0.6187573671340942, -0.20497170090675354, 0.17925187945365906, -0.1984584927558899, 0.29572975635528564, -1.0051119327545166, 0.08983872085809708, -0.10198816657066345, 0.3284505009651184, -0.19452179968357086, -0.171433687210083, -0.11427020281553268, -0.07383585721254349, -0.45427247881889343, 0.002701156074181199, 0.14893583953380585, -0.170203298330307, -0.23900432884693146]	0	0.618757	-0.204972	0.179252	-0.198458	0.295730	-1.005112	0.089839	...	0.328451	-0.194522	-0.171434	-0.114270	-0.
1	1	[0.8391729593276978, -0.21198837459087372, 0.1809975653886795, -0.30981990694999695, 0.3136947453022003, -1.1251457029611206, 0.08552251756191254, -0.19019414484500885, 0.35411620140075684, -0.19806264340877533, -0.1758519560098648, -0.06576484441757202, -0.08782333135604858, -0.564352810382843, -0.06888870894908905, 0.11816339194774628, -0.17591702938079834, -0.2600068747997284]	21	0.839173	-0.211988	0.180998	-0.309820	0.313695	-1.125146	0.085523	...	0.354116	-0.198063	-0.175852	-0.065765	-0.

Figure 7.9 – The res DataFrame after transformation: some columns are not visible in the screenshot

In order to get a better idea of what the embedding results are, we are going to plot the results. To do so, we are first going to reduce again the dimension of the feature space to only 2 so that we can plot it in a 2D scatter plot. We are also going to extract the community each node belongs to, using again the Louvain algorithm so that we can somehow see the graph topology and relate embedding encoding to it.

4. Compute Louvain communities:

```
louvain = gds.louvain.stream(graph_object)
```

Investigating the content of this DataFrame, you'll find out that four different communities are identified by this algorithm. Let's see if the graph embedding is also capable of capturing this information.

5. Append the Louvain `communityId` values to the main DataFrame. In order not to mix nodes, we are going to use `nodeId` as the index to make sure we assign the right community to each node (we do not have any guarantee that nodes are returned in the same order by every algorithm):

```
louvain_reindexed = louvain.set_index("nodeId")
res_reindexed = res.set_index("nodeId")
res_reindexed["louvain"] =
    louvain_reindexed.communityId
```

6. The Louvain `communityId` values are categorical numbers, whose values can be very different (from 1 to 400 while we only have four communities). In order to make the visualization nicer, we are going to encode them so that they use consecutive numbers:

```
from sklearn import preprocessing
le = preprocessing.LabelEncoder()
res_reindexed["louvain_encoded"] = \
   le.fit_transform(res_reindexed.louvain)
```

Note that this step is not mandatory; there are other ways to plot categorical variables, but we are doing so here to use grayscale, which is compliant with the printed version of this book.

7. We can then compute the **principal component analysis** (**PCA**) using the embedding columns to get a 2D representation of our nodes:

```
from sklearn.decomposition import PCA
pca = PCA(n_components=2)
embedding_pca = \    pca.fit_transform(res_reindexed.
loc[:,f_names])
```

8. Finally, we can plot the 2D `embedding_pca` variable on a scatter plot using `matplotlib`, each marker representing a node:

```
import matplotlib.pyplot as plt
plt.scatter(
   *zip(*embedding_pca),
```

```
        # marker color
         c=res_reindexed.louvain_encoded,
        cmap="gray",
        # marker size
        s=50
        # marker edge (line) color
        edgecolor="grey",
    )
```

The marker brightness is an indication of its Louvain community.

The resulting plot, after some more tuning to set axes labels and figure title, is reproduced as follows:

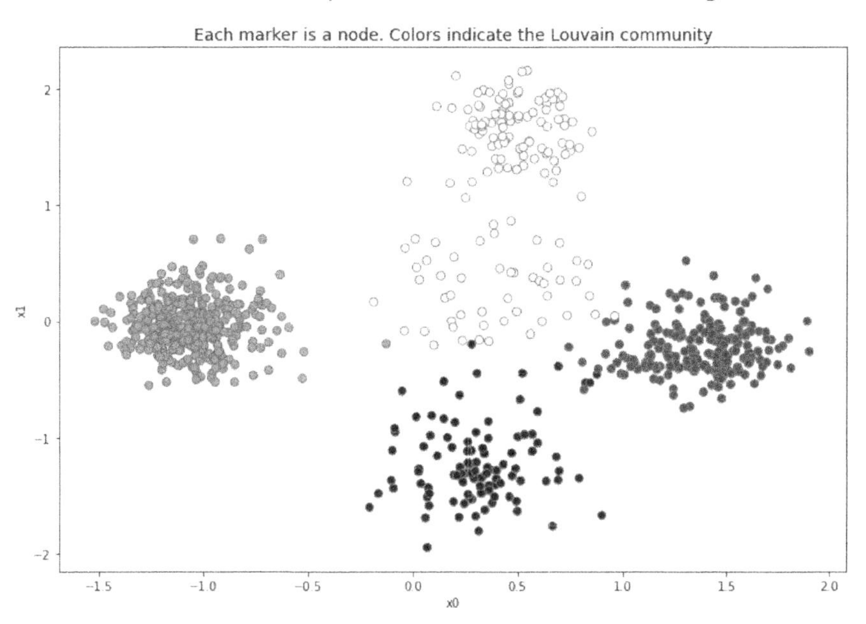

Figure 7.10 – 2D PCA decomposition of the node embeddings obtained with the Node2Vec algorithm

Randomness

Note that the Node2Vec algorithm is *non-deterministic*, meaning you will obtain slightly different results when running the code on your own.

As you can see, the Node2Vec algorithm has been able to nicely *capture the community structure of the graph*: nodes in the same community are closer to each other in the embedding space than to nodes in a different community (another shade of gray).

Despite this nice result, the Node2Vec algorithm suffers from a major drawback: it won't be able to predict the embedding of a newly added node since it would have to recompute random walks including this node and retrain the skip-gram model. Fortunately, other approaches exist. In the following section, we are going to learn about the **GraphSAGE** algorithm, which is able to predict the vector representation of unseen nodes.

Training an inductive embedding algorithm

GraphSAGE is another type of algorithm. Instead of learning the embeddings themselves, which prevents making predictions on new nodes, *it learns the function to compute the embeddings*, which, once learned, can be applied to unknown nodes. It also has the ability to take into account node properties, making it an interesting algorithm to mix the graph structure and node characteristics into one single vector. In this section, we are going to give some more details about GraphSAGE internals, before using it with our data stored in Neo4j.

Understanding GraphSAGE

GraphSAGE relies on the principle of *message propagation* in a graph, from one node to its neighbors, and aggregates the received information to iteratively build node representations. It is also known to be scalable due to its neighbor-sampling technique.

Message propagation

Using again the graph represented in *Figures 7.1* and *7.2*, we are first going to create a one-hot encoding representation for each node, a matrix of dimension *5x5* with 1s on the diagonal and 0s elsewhere:

```
n = 5
X = np.eye(n, n)
```

Then, we try to assign a lower-dimension vector to these nodes, using random weights:

```
embedding_dimension = 2
# initial random weights
W = np.random.uniform(-1, 1, (n, embedding_dimension))
```

The node positions according to the weight matrix `W` are illustrated on the left side of *Figure 7.8*.

If we want to propagate the weights to *node neighbors*, we have to use the adjacency matrix A:

```
A = np.array([
    [0, 1, 1, 1, 0],
    [1, 0, 0, 1, 1],
    [1, 0, 0, 1, 0],
    [1, 1, 0, 0, 0],
    [0, 1, 0, 0, 0],
])
```

And then, we perform a dot product between the weight matrix and this adjacency matrix:

```
layer0 = A.dot(W)
layer1 = A.dot(layer0)
...
```

For node A, for instance, we would have the following:

```
layer0_A = W_B + W_C + W_D
layer1_A = layer0_B + layer0_C + layer0_D
```

In that way, node A receives information from its direct neighbors at each iteration.

Repeating this operation several times, we would end up with the graph representation of the right-hand side of the following screenshot:

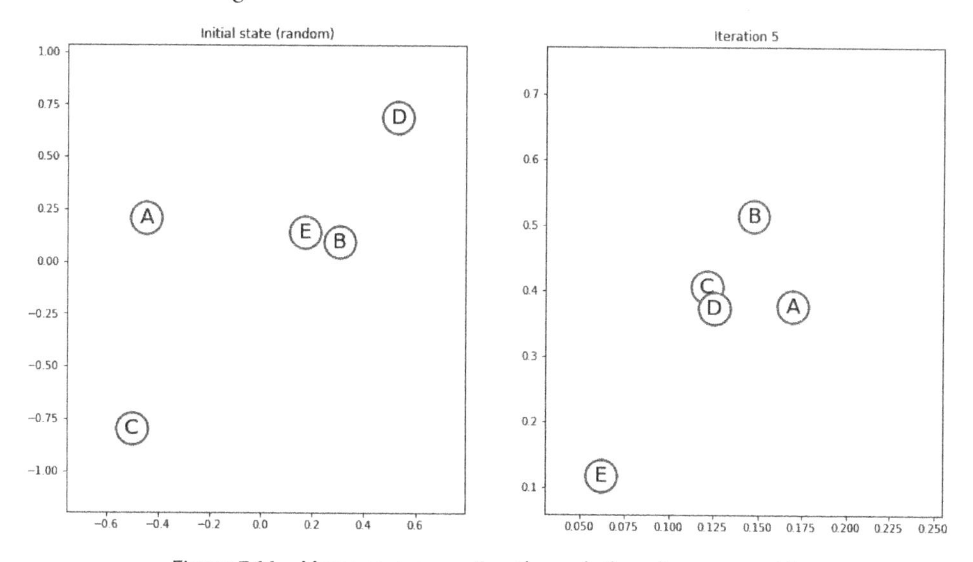

Figure 7.11 – Message propagation through the adjacency matrix

You can see that, from a totally random situation, we have somehow managed to identify that node **E** is different from the rest of the graph, being less connected to the other nodes.

This example is trivial and is not intended to be used as is; it's just intended to give you an intuition about the usefulness of message propagation through node neighbors. Message propagation is not specific to GraphSAGE and is not sufficient to efficiently compute meaningful and inductive embeddings. For this, we also need neighbor sampling and an aggregator function.

Neighbor sampling

GraphSAGE is known to be a scalable algorithm, able to perform well on quite large graphs (> 100,000 nodes for the GraphSAGE paper). This is achieved using neighbor sampling: instead of using all neighbors in the message-passing stage, a subset of all existing edges is selected. This allows us to keep the memory requirements relatively low.

Aggregator function

The aggregator function is the function that deals with all messages received from the neighbors and outputs a new embedding. In the previous example, the aggregator was just a sum:

```
layer1_A = layer0_B + layer0_C + layer0_D
         = sum(layer_0 of neighbors)
```

In practice, the aggregator function has some parameters, and these parameters are learned during the training phase of the algorithm so that when a new node is added to the graph, the algorithm just has to use the already computed embedding from the neighbors and aggregate them with the learned aggregator.

The particularity of GraphSAGE goes beyond these two steps. Indeed, it is designed for scalability and performs efficient neighbors sampling, in order to reduce the aggregation complexity on very large graphs.

Let's now use the GraphSAGE embedding algorithm with the GDS library.

Introducing the GDS model catalog

We already know from previous chapters that GDS is able to store a projected graph in memory. But this is not the only kind of object GDS can store. For a model such as GraphSAGE that can be trained and reused as new observations arrive, the trained model can be stored in a **model catalog**. It works pretty much like the graph catalog. Models are uniquely identified by their names and can be dropped at any time when they are not needed anymore.

To list all available models, you'll use the following function:

```
gds.beta.model.list()
```

Dropping a model not used anymore is performed with the `drop` function:

```
gds.beta.model.drop(model)
```

Note

Similarly to the graph catalog, while the Cypher procedures expect the model name as textual input, the equivalent functions in the Python driver expect a `model` object.

Notice the `beta` prefix in the function name.

Beta version

At the time of writing, the model catalog is still a beta feature of GDS (version 2.2). This may evolve in the future, and it is very likely that in a future version, this feature will be promoted, and hence the `beta` prefix for model-related functions will be dropped.

Right now, we know how to list models in the model catalog and drop an existing model, but we still can't create a model. There is no such function as `gds.model.create`. In order to add a model to the model catalog, we need to *train* it. That's what we are going to do by training a GraphSAGE embedding algorithm.

Training GraphSAGE with GDS

In order to create a new model, let's go ahead and train it. In order to learn both the training and prediction stages, we are going to create training and testing sets. Nodes in the training set will be assigned a `TRAIN` label, while nodes in the testing set will have a `TEST` label. All the nodes currently in our graph (with a `NodeB` label) will be in the training set, but we will add some new nodes to it for our testing set. These steps are performed in the `GraphSAGE_Train_Predict` notebook. We will discuss them here:

1. Assign some nodes another label, `Test`:

   ```
   MATCH (n:Node)
   WITH rand() as r, n
   WHERE r < 0.2
   SET n:Test
   ```

2. And assign the remaining nodes a `Train` label:

   ```
   MATCH (n:Node)
   WHERE NOT n:Test
   SET n:Train
   ```

3. Import the `x0` node feature:

```
LOAD CSV WITH HEADERS FROM "file:///x0.csv" AS row
MATCH (n:Node {id: toInteger(row.id)})
SET n.x0 = toFloat(row.x0)
```

4. Create a projected graph with all nodes with a `Train` label:

```
from utils import create_projected_graph

projected_graph_object = create_projected_graph(
    gds,
    graph_name="pGraphAll",
    node_spec=["Train"],
    relationship_spec={"RELATED_TO": {"orientation":
"UNDIRECTED"}},
    nodeProperties=["x0"],
)
```

Note that we are also adding to the projected graph the `x0` node property.

5. Run the GraphSAGE algorithm on the projected graph, filtering nodes using the `Train` label, and assign the name `myGS` to the stored model:

```
model_object, res = gds.beta.graphSage.train(
    graph_object,
    modelName="myGS",
    featureProperties=["x0"],
    learningRate=0.0001
)
```

We are also telling GraphSAGE to use `x0` as an initial feature and tuning the learning rate so that the algorithm converges and finds meaningful results.

Similarly to what we have done with projected graphs, we have to store the `model` object here to reuse it later on when we want to make predictions based on this model.

Now, we can list the models in the catalog and see our newly trained model called `myGS`, as illustrated in the following screenshot:

```
gds.beta.model.list()
```

	modelInfo	trainConfig	graphSchema	loaded	stored	creationTime	shared
0	{'modelName': 'myGS', 'modelType': 'graphSage'...	{'maxIterations': 10, 'negativeSampleWeight': ...	{'graphProperties': {}, 'relationships': {'LIN...	True	False	2022-09-04T15:39:27.931282000+02:00	False

Figure 7.12 – Available models in the model catalog

We can also investigate the content of the `res` variable: it contains information about the trained model and its configuration, as illustrated in the following screenshot:

```
: res["modelInfo"]

{'modelName': 'myGS',
 'modelType': 'graphSage',
 'metrics': {'ranIterationsPerEpoch': [2],
  'iterationLossesPerEpoch': [[26.578491088885535, 26.57849037081301]],
  'didConverge': True,
  'ranEpochs': 1,
  'epochLosses': [26.57849037081301]}}

: res["configuration"]

{'maxIterations': 10,
 'negativeSampleWeight': 20,
 'searchDepth': 5,
 'aggregator': 'MEAN',
 'activationFunction': 'SIGMOID',
 'penaltyL2': 0.0,
 'learningRate': 0.0001,
 'concurrency': 4,
 'jobId': '77ae913d-3cf8-4ea5-831f-317e4ac711f3',
 'modelName': 'myGS',
 'embeddingDimension': 64,
 'nodeLabels': ['*'],
 'sudo': False,
 'featureProperties': ['x0'],
 'sampleSizes': [25, 10],
 'relationshipTypes': ['*'],
 'batchSize': 100,
 'epochs': 1,
 'tolerance': 0.0001}
```

Figure 7.13 – Model metrics after training and configuration

> **Model storage**
>
> The model catalog stores models in memory. It is, however, worth mentioning that it's possible to *persist the models on disk* and reload them later on, using the functions in the alpha tier at the time of writing: `gds.alpha.model.store(<model>)` and `gds.alpha.model.load(<model>)`.

It is now time to use this trained model to make predictions on unseen nodes.

Computing new node representations

Similarly to the previously studied algorithms, we can extract the results via streaming, writing to Neo4j, or mutating the projected graph. We will learn about the use case for the later mode in the next chapter (*Chapter 8, Building a GDS Pipeline for Node Classification Model Training*), but for now, let's use the simple streaming mode:

```
graph_object = create_projected_graph(
    gds,
```

```
        "pgraphTest",
        node_spec="Test",
        relationship_spec={
            "RELATED_TO": {"orientation": "UNDIRECTED"}
        },
        nodeProperties=["x0"]
)

embeddings = gds.beta.graphSage.stream(
        graph_object,
        modelName="myGS",
        nodeLabels=["Test"]
)
```

As usual, we need to provide the projected graph object. Since we want to make predictions from a given model, we also need to specify which model to use through the modelName configuration parameter.

The preceding code is equivalent to this one, where we use a predict method on the model object to obtain the embeddings:

```
embeddings = model_object.predict_stream(
        graph_object,
)
```

In both cases, the embeddings variable is a DataFrame containing one row per node with a Test label and two columns: nodeId (the GDS internal node ID) and embedding. Similarly to the Node2Vec results, embedding is a list with embeddingDimension elements (the default for GraphSAGE is 64).

> **Input features**
>
> As you may expect, the new nodes need to have the same features as the nodes used during training. In our case, that means that nodes with a Test label need to have an x0 property.

This was the final remark in this introductory chapter about node embeddings. There is much more to say about it, but it would require a full book. There are already very well-written books on this topic. I'll list some of them I had the opportunity to read in the *Further reading* section at the end of this chapter for those of you who want to investigate this topic further. Graph ML is definitely of great interest to data scientists today. It has already proven to be a valuable tool and will very likely continue to be with the new progress that is made every day in this area of knowledge.

Summary

In this chapter, you have learned about graph embeddings, first learning what embedding is and that nodes, edges, and whole graphs can be vectorized independently. Focusing on node embeddings, you have then learned the principles of two algorithms included in the GDS library. Node2Vec, inspired by the world of texts, is a transductive algorithm, while GraphSAGE, a message-passing algorithm, is inductive and is able to predict the embedding of unseen nodes.

You have been able to extract node embeddings for nodes stored in Neo4j using the GDS implementation of these algorithms. In addition, you have discovered the GDS model catalog and been able to train a GraphSAGE model, store it into the GDS in-memory model catalog, and reuse it to predict new node representations.

In the coming chapters, we will use these embedding models and discover a new feature of GDS: **pipelines**. We will use these pipelines to train a node classification model. In the following chapter, we will start this exploration to revisit node classification, and we will build a GDS pipeline that will take care of the whole process, from feature selection to model training.

Further reading

If you want to learn more about graph embeddings, I recommend the following readings:

- *Graph Machine Learning* by *C. Stamile, A. Marzullo,*and *E. Deusebio, Packt Publishing*. It's a comprehensive introduction to Graph ML. Both supervised and unsupervised algorithms are covered, with applications in various fields including **natural language processing** (**NLP**), using `networkx` and Python ML libraries such as `tensorflow`. It is a nice complement to this book. As an exercise, you can try to redo the analyses presented in the GML book using the tools we are discussing in this book, Neo4j and GDS.
- P-GNN algorithm—I talked about this in the *Positional or structural* section: `https://snap.stanford.edu/pgnn/`
- I presented some graph embedding algorithms in this *Medium* story: `https://medium.com/@st3llasia/graph-embedding-techniques-7d5386c88c5`
- The original DeepWalk paper: `https://arxiv.org/abs/1403.6652`
- The original Node2Vec paper: `https://arxiv.org/abs/1607.00653`
- *Complete guide to understanding Node2Vec algorithm*, by *T. Bratanic*, presents in more detail the algorithm parameters and how to choose them: `https://towardsdatascience.com/complete-guide-to-understanding-node2vec-algorithm-4e9a35e5d147`
- GraphSAGE explanations by its authors (`https://snap.stanford.edu/graphsage/`) and related paper: `https://arxiv.org/abs/1706.02216`

- A nice paper on how to measure the performance of a graph embedding algorithm: `http://www.doiserbia.nb.rs/img/doi/1820-0214/2019/1820-02141900011L.pdf`
- Also about embedding *correctness*, this story by N. Smith, data scientist at Neo4j, is a must-read: `https://towardsdatascience.com/understanding-graph-embeddings-with-neo4j-and-emblaze-7e2d6ef56b8c`

Exercises

To test your understanding, try to answer the following questions:

1. Can you give two reasons why we need embedding techniques?
2. What is the drawback of reducing the dimension of a dataset?
3. What kind of information can be encoded from a graph structure?
4. Would you use a transductive algorithm to make predictions on unseen nodes?
5. What are the embedding algorithms implemented in the GDS library?
6. What are the advantages of GraphSAGE over Node2Vec?
7. Where are the inductive algorithms stored after training in the GDS library? Why do we need to store them?

8

Building a GDS Pipeline for Node Classification Model Training

Classifying observations within categories is a classical **machine learning** (**ML**) task. As we learned in the preceding chapters, we can use existing ML models such as decision trees to classify a graph's nodes. The graph structure is used to find extra features, bringing more knowledge into the model. In this chapter, we will discover another key feature of the Neo4j GDS library: pipelines. They let you configure and train an ML model, before using it to make predictions on unseen nodes. You can do all of this from Neo4j, without having to add another library such as scikit-learn to the tech stack.

Also, we are going to work on the Netflix dataset we created earlier in this book (the code is available on GitHub if you don't have it yet). We will try and make predictions by building a node classification pipeline, focusing on the how rather than the why.

In this chapter, we're going to cover the following main topics:

- The GDS pipelines
- Building and training a pipeline
- Making predictions
- Using embedding features

Technical requirements

In order to be able to reproduce the examples given in this chapter, you'll need the following tools:

- Neo4j 5.x installed on your computer (see the installation instructions from *Chapter 1, Introducing and Installing Neo4j*):
  - The Graph Data Science plugin (version >= 2.2)

- A Python environment with the following:
  - Jupyter to run the notebooks
  - `scikit-learn`
- Any code listed in the book will be available in the associated GitHub repository (`https://github.com/PacktPublishing/Graph-Data-Science-with-Neo4j`) in the corresponding chapter folder

> **Code samples**
>
> *Unless otherwise indicated, all code snippets in this chapter and the following ones use the GDS Python client.* Library import and client initialization are omitted in this chapter for brevity, but a detailed explanation can be found in the *Introducing the GDS Python client* section *of Chapter 6, Building a Machine Learning Model with Graph Features*. Also, note that the code in the code bundle provided with the book is fully runnable and contains all imports.

The GDS pipelines

This section introduces GDS pipelines, where we explain what the purpose of this feature is, illustrate its intended usage, and show the basic usage of the pipeline catalog.

What is a pipeline?

As data scientists, we run data pipelines every day. Any logical flow of action is somehow a pipeline, and when you run your Jupyter notebook, you already have a pipeline. However, here, we refer to explicitly defined workflows, with sequential tasks such as the one we can build with `scikit-learn`. Let's take a look at the `Pipeline` object in this library before focusing on GDS pipelines to understand their similarities and differences.

scikit-learn pipeline

Often, we think about ML as finding the best model for a given problem, but as data professionals, we know that finding the right model is only a small part of the problem. Before we can even think about fitting a model, many preliminary steps are required: from data gathering to feature extraction. Some steps cannot be automated with code; for instance, when you need to talk to the field expert to understand how the data was collected. But once you have a good-enough understanding of the meaning of each column in your dataset, you can start cleaning the data and preparing it for a given model. That includes dealing with outliers, missing data, sometimes data scaling, and more. At that stage, you can easily end up with a messy notebook that includes steps like the following:

```
data["x1"] =
data["x2"] = data["x2"].fillna(0)
```

```
data["x3"] = ...

scaler = StandardScaler().fit(X_train)
model =  LinearSVC().fit(X_train)
```

During experimentation, often, we have this code split into multiple cells in a Jupyter notebook, not always executed in the order they appear on the screen. Moreover, we almost always need to repeat some steps between the training steps and the testing steps (for instance, for data scaling). Hence, we have to repeat a lot of code between training and testing. To avoid this, you can use the `scikit-learn` pipeline instead. The following code snippet shows you how to create such a pipeline:

```
pipe = Pipeline([
     # feature transformation
     ("scaler", StandardScaler()),
     # classification model
     ("classifier", LinearSVC())
])
```

It requires a list of steps, each made of a name and a transformer object that can be applied to the data. The next step is to fit the pipeline to the training data with the following:

```
pipe.fit(X_train, y_train)
```

This simple line takes care of fitting all the transformers in the pipeline. In our example, both the scaler and the classifier will be fitted on the training data and stored in the pipeline object. Then, making predictions is as simple as the following:

```
pipe.predict(X_test)
```

We can also compute the model performance with the following:

```
pipe.score(X_test, y_test)
```

As you will see, GDS pipelines are similar, but there are some key points to remember.

The GDS pipeline lifecycle

The Neo4j Graph Data Science library also contains the concept of a **pipeline**, in the context of a graph. It shares similar concerns, such as feature extraction and model training. Its lifecycle is illustrated in the following diagram:

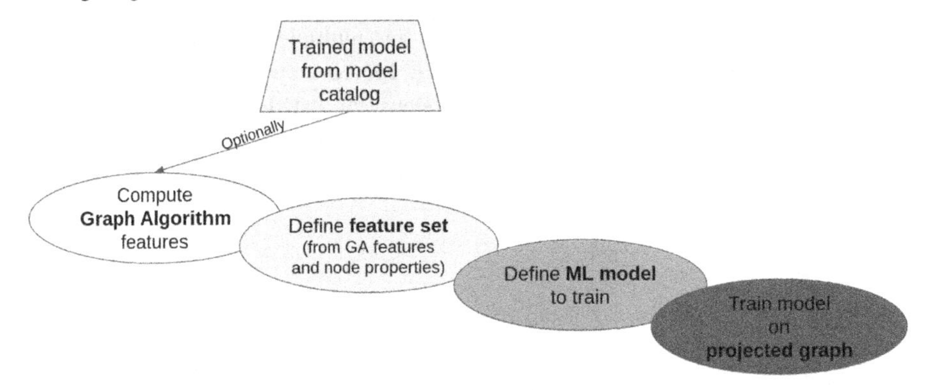

Figure 8.1 – GDS pipeline training steps

We will understand each of these steps as we proceed with an example in the following section. First, let me introduce the pipeline catalog.

The GDS pipeline catalog

Similarly to projected graphs (see *Chapter 4, Using Graph Algorithms to Characterize a Graph Dataset*) and models (see *Chapter 7, Automatically Extracting Features with Graph Embeddings for Machine Learning*), GDS pipelines (from now on, referred to as pipelines) are stored in a *catalog*. Similar functions exist to check the catalog content and retrieve or delete a specific object:

```
print(gds.beta.pipeline.list())
pipeline_obj = gds.pipeline.get("<pipelineName>")
gds.beta.pipeline.drop(pipeline_obj)
```

With these three lines, we are doing the following:

1. Listing all the existing pipelines. This returns a DataFrame containing several columns, along with `pipelineName`, which we use in the next line.
2. With the `gds.pipeline.get` function, we retrieve the pipeline object from its name (string).
3. This pipeline object is then used as input for the `gds.beta.pipeline.drop` function to remove this pipeline from the store.

This is how you can clean your pipeline catalog from Python. Note that, as with projected graphs and models, the pipeline catalog is in-memory, meaning it won't survive a Neo4j instance restart.

Note about the GDS version

In GDS version 2.2, functions related to the pipelines are still in the beta tier, but they are likely to be promoted in a coming release, which means the beta prefix will likely disappear. You can use the following code to find out the function name for your version of the library:

```
functions = gds.list()
functions [functions.name.str.contains("pipeline")]
```

Now, let's see something more interesting: how to create, train and, make predictions with a pipeline.

Building and training a pipeline

Similarly to models, in order to add a pipeline to the catalog, we'll have to train it. Pipeline training requires several steps:

1. Create and name the pipeline object.
2. Optionally, compute features from other GDS algorithms (such as graph algorithms, embeddings, or pre-processing).
3. Define the *feature set* from the features added in the previous step, and/or any node property included in the projected graph.
4. Select the ML models to be tested with their hyperparameters: The pipeline training will run all algorithms and select the best one.
5. Finally, train the model.

The following sub-sections detail each of these steps. The supporting notebook is `Pipeline_Train_Predict`. This can be found in the `Chapter08` folder of the code bundle that comes with this book.

Creating the pipeline and choosing the features

In GDS, we can create *three* kinds of pipelines:

- **Node classification**: Each node gets assigned to one target class in a supervised manner.
- **Node regression**: The model is trying to predict the value of a numeric node property (float).
- **Link prediction**: The goal of this kind of model is to predict future or hidden links in a partial or time-evolving network. We will come to this use case in *Chapter 9, Predicting Future Edges*.

The problem we are trying to solve

In this chapter, we are trying to solve the same problem we studied in *Chapter 6, Building a Machine Learning Model with Graph Features*. This is a node classification problem using the person identified in the Netflix dataset that we created in *Chapter 2, Using Existing Data to Build a Knowledge Graph*, trying to predict whether these persons are US citizens or not. If you need to import the data, please follow the instructions in the `README` file for this chapter, which you'll find in the GitHub repository or code bundle for this book.

The code for this section is available in the `GDS_Pipeline` notebook.

So, our problem is a node classification problem; hence, we can create our classification pipeline using the `nodeClassification` function like this:

```
pipeline_object, res =  gds.beta.pipeline\
   .nodeClassification\
   .create("nodeClassificationPipeline")
```

The `res` variable contains interesting information about the pipeline configuration. Just after its creation, it shows the default parameters illustrated in the following screenshot:

```
name                                       nodeClassificationPipeline
nodePropertySteps                                                  []
featureProperties                                                  []
splitConfig               {'testFraction': 0.3, 'validationFolds': 3}
autoTuningConfig                                    {'maxTrials': 10}
parameterSpace      {'RandomForest': [], 'LogisticRegression': []}
Name: 0, dtype: object
```

Figure 8.2 – Default pipeline configuration

This configuration will evolve as we tune our pipeline, which we are going to do next.

Setting the pipeline configuration

When defining a pipeline, we need to define the features to be used by our model, the type of model to use, and its hyperparameters. That's what we will do in the following sub-sections, starting with the ability to add new node properties in the projected graph.

Adding node properties – graph algorithm features

GDS pipelines have the ability to add properties to nodes in the projected graph, through another GDS algorithm such as node importance (e.g., PageRank) or community detection (e.g., Louvain) algorithms. Additionally, it is possible to apply some preprocessing algorithms, such as scaler. To do this, we would need to use the `addNodeProperty` function, where the first parameter is the name of the procedure to be called. We also need to provide the parameters needed by the algorithm procedure, such as the name of the property that will be written in the projected graph (`mutateProperty`). For instance, let's add to our pipeline the Louvain algorithm. For this, we need the following:

```
pipe.addNodeProperty(
     "gds.louvain.mutate",
     mutateProperty="lv"
)
```

The `gds` prefix and the `mutate` suffix can actually be omitted, which simplifies the notation. The preceding code is equivalent to the following:

```
pipe.addNodeProperty("louvain", mutateProperty="lv")
```

However, due to reasons we will tackle in the next section, we are going to use the Louvain algorithm through another method.

Preparing the data

To simplify things a little bit, we are also going to add properties to our Neo4j database. Then, we will be able to do some cleaning, before starting to build our model:

1. First, the number of movies each person is involved in, either as an actor or a director, that we've used as a feature in *Chapter 6, Building a Machine Learning Model with Graph Features*:

   ```
   MATCH (p:Person:MainComponent)-[:ACTED_IN|DIRECTED]-
   >(m:Movie)
   WITH p, count(DISTINCT m) as nbMovies
   SET p.nbMovies = nbMovies
   ```

2. We are also going to add an `isUSCitizen` property, which will be our target:

   ```
   MATCH (p:Person:MainComponent)-[:IS_CITIZEN_
   OF]->(c:Country)
   SET p.isUSCitizen = toInteger(
        c.name = "United States of America"
   )
   ```

> **Note**
>
> These additions are not necessary, as these properties could be created at runtime when projecting the graph with a Cypher projection. I'll leave it to you as an exercise to try and build the projected graph with this kind of projection.

3. Next, let's clean our data, and remove the nodes that do not have the `isUSCitizen` property (the call to Wikidata to find their country of citizenship failed):

```
MATCH (p:MainComponent)
WHERE p.isUSCitizen IS NULL
REMOVE p:MainComponent
```

4. We then set a new label in the database to set aside some nodes for testing:

```
MATCH (p:Person:MainComponent)
WITH p, rand() as r
WHERE r > 0.1
SET p:MainTrain
```

5. Next, we set aside the corresponding test nodes:

```
MATCH (p:Person:MainComponent)
WHERE NOT p:MainTrain
SET p:MainTest
```

6. Finally, we will randomly sample some nodes from the train sample so that we reduce the class imbalance between US and non-US citizens. First, we add all US citizens to this new train set, returning the total count:

```
MATCH (p:Person:MainTrain)
WHERE toBoolean(p.isUSCitizen) = true
SET p:MainTrainDownSampled
WITH count(p) as nbUSCitizen
RETURN nbUSCitizen
```

7. We end up with `6872` nodes. Then, we randomly select some non-US citizens up to `6872`:

```
MATCH (p:Person:MainTrain)
WHERE toBoolean(p.isUSCitizen) = false
WITH p ORDER BY rand() LIMIT 6872
SET p:MainTrainDownSampled
```

And that's it.

Let's recap since we have created quite a lot of labels. Our graph contains the following labels:

- `MainComponent`: This includes all nodes in the main component (connected component-wise) and with the `isUSCitizen` property set

- `MainTest`: This is about 10% of the nodes in the `MainComponent` set
- `MainTrain`: This includes nodes in the `MainComponent` set that are not in the test set
- `MainTrainDownSampled`: This includes nodes from the `MainTrain` set so that US and non-US citizens are represented evenly

Due to the random downsampling of the non-US citizens, we are not going to compute the Louvain property within the pipeline, as it will be biased. Instead, we are going to compute the community structure on the whole main component and save the result in a node's property thanks to the `write` execution mode of the algorithm:

```
CALL gds.graph.project(
     "pgraph",
     "MainComponent",
     {KNOWS: {orientation: "UNDIRECTED"}}
)
CALL gds.louvain.write("pgraph",
     {writeProperty: "louvain"})
```

Even if we have not used it for sampling the dataset, you can check that the `louvain` feature is quite evenly distributed between different sets (train versus the downsampled train, and train versus the test dataset).

Now it is time to select the features we want our model to be trained on.

Adding node features – defining the features set

The features must be selected from the added node properties of the previous step and the properties included in the projected graph when creating it, since models are trained on a projected graph, like anything else in the GDS. So, let's create a projected graph.

With these graph updates, we can proceed and create a projection (here, we are using the Python client, but it could also be done within Neo4j Browser):

```
from utils import create_projected_graph
pgraph_train = create_projected_graph(
     gds,
     graph_name="graph-clf",
     node_spec={
          "Train": {
               "label": "MainTrainDownSampled",
               "properties": {
                    "nbMovies": {"defaultValue": 0},
```

```
                    "louvain": {},
                    "isUSCitizen": {},
                }
            },
            "Test": {
                "label": "MainTest",
                "properties": {
                    "nbMovies": {"defaultValue": 0},
                    "louvain": {},
                    "isUSCitizen": {}
                }
            }
        },
        relationship_spec={
            "KNOWS": {
                "orientation": "UNDIRECTED",
                "aggregation": "SINGLE"
            }
        },
)
```

The `create_projected_graph` function is a small wrapper around the `gds.graph.project` procedure that will check whether a projected graph with the same name does not already exist and drop it otherwise. The `utils` module is available in the code bundle you should have received with this book.

Our projected graph will be made of the following:

- Nodes with the `Train` label, which are fetched from the `MainComponentDownSampled` nodes in the database.

  From these nodes, we copy the `isUSCitizen`, `louvain`, and `nbMovies` properties into the projected graph.

- Nodes with the `Test` label, which correspond to nodes with the `MainTest` label in Neo4j. We also copy the same properties. Note that `isUSCitizen` is usually not required for the test set since we are not supposed to know the target class during the prediction phase, but it will be used to test model performances.

- Relationships between these nodes with the `KNOWS` type.

  Note that, since we are not using a graph algorithm within the pipeline, the relationships are, in practice, not used, but they are required to create a projected graph.

At this stage, the important thing is that we include, in the projected graph, both the properties to be used as features (`nbMovies`), along with the target property (`isUSCitizen`).

Now we can go ahead and continue our pipeline configuration by defining the features:

```
pipe.selectFeatures(["nbMovies", "louvain"])
```

If you look at the pipeline object now, you'll see that its configuration has changed. In particular, it contains our feature definition:

```
...
'featurePipeline': {
  'nodePropertySteps': [],
  'featureProperties': [
        {'feature': 'nbMovies'}, {'feature': 'louvain'}
     ]
}
...
```

Now we are ready for the next step: defining the model to be used.

Setting the model

The GDS supports several classification models:

- Logistic regression
- Random forest
- **Multi-Layer Perceptron** (**MLP**)

Each of these is added with the `addX` function, where *X* is either `LogisticRegression`, `RandomForest`, or `MLP`. In our case, we are only going to use the random forest algorithm, so we just need to run the following code:

```
pipe.addRandomForest(minSplitSize=30, minLeafSize=10)
```

If we want to test different algorithms and/or different hyperparameter configurations, we just need to add multiple configurations; for instance, we could add `pipe.addLogisticRegression()` and `pipe.addRandomForest(maxDepth=20)`. Then, the pipeline would find the best one and use it for prediction. But here, let's keep it simple and continue with one single random forest.

Our pipeline now contains all the required information to be trained! We will start training it in the next section.

Training the pipeline

Once our pipeline has been configured, the only thing left is to actually train it. We do this by calling the train method of the pipeline object and specifying the following:

- The *projected graph* to use as input (where the model will find the node information): The projected graph must contain the selected node features and target property
- The *name of the model* that will be created for us and stored in the model catalog
- The *target node labels* to use for training
- The *target property* (the node property we are trying to predict)

Here is how to train our pipeline:

```
model_object, train_result = pipeline_object.train(
    projected_graph_object,
    modelName="model-clf-rf",
    targetNodeLabels=['Train'],
    targetProperty="isUSCitizen",
    metrics=["PRECISION(class=1)", "ACCURACY"],
)
```

The preceding returns both a `model_object` and some information about the training phase in a `train_result` `pd.Series`.

Wrap-up

When dealing with pipelines, we need to use projected graphs, models, and pipelines:

- The pipeline itself contains all the configurations. It will automatically run graph algorithms, if required, by the `addNodeProperty` function.
- Its *input dataset* is the data contained in a given *projected graph*. The algorithm results are stored in the projected graph thanks to the `mutate` function (this is done internally; the user does not have anything to do here).

  Note that the properties created in a projected graph during pipeline training are automatically dropped once the training is finished.
- It *outputs* a new *trained model*, stored in the model catalog, that can be used to make predictions.

This last bullet is the topic of the coming section.

Making predictions

In order to make predictions, we are going to use the same projected graph that already contains the test nodes.

With this projected graph, and the model object returned by the pipeline training, we can now predict the class of new nodes:

```
predictions = model.predict_stream(
      projected_graph_object,
      targetNodeLabels=["Test", "Train"],
)
```

Note that the model object also exposes a `predict_mutate` function to store the results in the projected graph. This will be useful to us when dealing with embedding features in the last section of this chapter.

In the preceding code block, we include both the `Test` and `Train` nodes in order for the Louvain results to be computed properly, using the whole graph. We will filter out the predictions for the train nodes as we evaluate the model performances.

For instance, in order to evaluate our model, we can compute the confusion matrix using our preferred tool. In the next section, an example is given using `scikit-learn` tools.

Computing the confusion matrix

Let's investigate the model result by looking at how many of its predicted classes are actually correct:

1. First, let's fetch the real classes. There are different ways of doing this. We are going to read the property from the projected graph directly:

   ```
   df_test = gds.graph.nodeProperty.stream(
         projected_graph_object,
         "isUSCitizen"
   )
   ```

   Here, `df_test` contains two columns: the GDS internal `nodeId` column and a `propertyValue` column.

2. Then, we set the index of both dataframes to `nodeId` so that we can more easily join them:

   ```
   df_test = df_test.set_index("nodeId")
   predictions = predictions.set_index("nodeId")
   ```

3. We can then add the predictions to the dataframe:

```
df_test["prediction"] = predictions.predictedClass
```

4. Then, we need to filter out nodes from the train sample, which we can achieve with the following lines:

```
test_nodes = ["MainTest" in n.labels for n in gds.util.
asNodes(list(df_test.index))]
df_test = df_test[test_nodes]
```

5. Now we can compute the confusion matrix with the following:

```
from sklearn.metrics import confusion_matrix
mat = confusion_matrix(
        df_test.propertyValue,
        df_test.prediction
)
```

The confusion matrix is as follows (you might get slightly different results due to random samplings):

```
array([[1045,  313],
       [ 517,  170]])
```

This means that among the non-US citizens, the pipeline correctly classified `1045` nodes and misclassified `313` nodes. However, the performances for the US citizen are not great: only 25% of them are classified properly. We could try and tune our model, but we are going to directly switch to using another type of feature: embeddings.

Last but not least, we are going to learn how to combine the pipelines we just studied with the node embedding algorithms we covered in the preceding chapter (*Chapter 7, Automatically Extracting Features with Graph Embeddings for Machine Learning*).

Using embedding features

The performed analysis is equivalent to the analysis performed in *Chapter 6, Building a Machine Learning Model with Graph Features*, with `scikit-learn`, except that here, there is no need to add another package for model training, as everything is taken care of in GDS.

However, in the preceding chapter, we learned about another way to find node features, by learning them from the graph structure itself: node embeddings. In this section, we will use node embeddings as features for our classification task.

Choosing the graph embedding algorithm to use

In *Chapter 7, Automatically Extracting Features with Graph Embeddings for Machine Learning*, we talked about two graph embedding algorithms included in GDS: Node2Vec and GraphSAGE. They have some differences, and one of them is the kind of information they tend to encode. While Node2Vec tends to model the node positions in the graph (nodes close to each other in the graph will have close embeddings), GraphSAGE is better at modeling the graph structure (nodes with similar neighborhood topology will have similar embeddings).

For our purpose, predicting whether a person is a US citizen or not, we can consider that the positional embeddings will bring more interesting information: persons close to each other in the graph have collaborated more than others and are more likely to have the same citizenship. We will start by using the Node2Vec algorithm.

Training using Node2Vec

Training using the Node2Vec algorithm is very similar to what we have done before with the Louvain algorithm. I'll just highlight the changes in this section, but the full code is available in the `GDS_Pipeline_Embedding_Node2Vec` notebook:

1. First, let's create a projected graph on which to compute the embeddings:

```
projected_graph_object = create_projected_graph(
     gds,
     graph_name="graph-clf-emb-n2v",
     node_spec="MainComponent",
     relationship_spec={"KNOWS": {"orientation":
"UNDIRECTED", "aggregation": "SINGLE"}},
)
```

2. Next, run the Node2Vec algorithm and store the results back in Neo4j:

```
gds.beta.node2vec.write(projected_graph_object,
writeProperty="n2vEmbedding")
```

3. From there, the process is very similar to what we have done in the previous section. First, we create a new projected graph containing the embedding property:

```
projected_graph_object = create_projected_graph(
     gds,
     graph_name="graph-clf-emb",
     node_spec={
           "Person": {
```

```
                "label": "MainTrainDownSampled",
                "properties": {
                        "isUSCitizen": {"defaultValue":
0},
                        "n2vEmbedding": {},
                }
            },
        },
        relationship_spec={"KNOWS": {"orientation":
"UNDIRECTED", "aggregation": "SINGLE"}},
)
```

4. Then, we create and configure the pipeline:

```
pipe, pipe_info = gds.beta.pipeline.nodeClassification.
create("pipe-clf-emb")
pipe.selectFeatures(["n2vEmbedding"])
pipe.addRandomForest(minSplitSize=30, minLeafSize=10)
```

5. Once the configuration is done, we can proceed with pipeline training;

```
model, model_info = pipe.train(
     projected_graph_object,
     modelName="clf-rf-emb",
     targetProperty="isUSCitizen",
     metrics=["PRECISION(class=1)", "F1_WEIGHTED",
"ACCURACY"],
)
```

The prediction and model evaluation parts are strictly identical, and they yield the following results:

```
array([[1125,  233],
       [  61,  626]])
```

This time, our classification performs much better, misclassifying only 394 (233 + 61) observations out of 2,000. The confusion matrix is reproduced in the following screenshot:

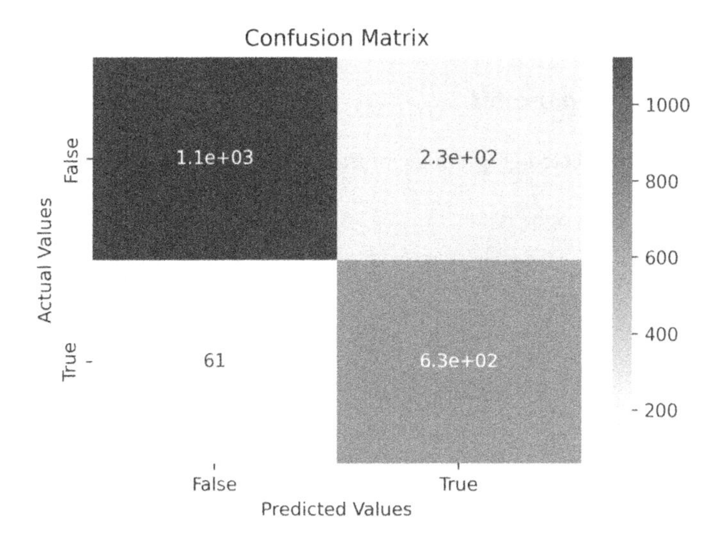

Figure 8.3 – Classification confusion matrix using the Louvain feature

However, these results are questionable. Since we've trained the Node2Vec algorithm on the full dataset (train + test), we have some data leakage from the test set to the train set. This is because Node2Vec is built from random walks through the whole graph, *including test nodes*. Somehow, the train set is already aware of its neighborhood. If we trained the algorithm once on the train set, trained the model, and then retrained Node2Vec on the full dataset to make predictions, we'd have ended up with unrelated embeddings, Node2Vec being undeterministic even on the same dataset, and the predictions would fail because the test observations would span a different vector space than to the train ones.

After this short discussion, let's see how to leverage the inductive nature of GraphSAGE for the same purpose of node classification. Even if the embeddings are not encoding the information we are interested in and we do not expect the model to perform well, it is interesting for us to learn how the process works with the GDS.

Training using GraphSAGE

The structure of the pipeline is exactly the same. The only difference, and what makes GDS pipelines different from `scikit-learn` ones, is that the embedding model *needs to be trained before being included in the pipeline*. The pipeline will call the `predict_mutate` procedure for the embedding model to extract the embeddings and use them as features. This also means we have to be careful about our train/test splits, as we do not want data leakage between the embedding training and the classification model training. In this section, we will consider whether we can use the same training set for the embedding algorithm and the classifier, which will both use the `MainTrain` labels (or the subset `MainTestDownSampled` label) for training.

The code for this section is available in the `GDS_Pipeline_Embedding_GraphSAGZ` notebook.

Training the embedding model

We are going to create a projected graph made of train nodes and train a GraphSage model on it:

1. Create the projected graph:

```
projected_graph_object = create_projected_graph(
    gds,
    graph_name="graph-clf-emb-gs",
    node_spec={
            "Person": {
                    "label": "MainTrain",
                    "properties": {
                            "nbMovies": {"defaultValue": 0},
                    }
            }
    },
    relationship_spec={"KNOWS": {"orientation":
"UNDIRECTED", "aggregation": "SINGLE"}},
)
```

2. Train the GraphSAGE node embedding:

```
gs_model, _ = gds.beta.graphSage.train(
    projected_graph_object,
    modelName="model-gs-clf-emb",
    featureProperties=["nbMovies"],
    learningRate=0.0001
)
```

The GraphSAGE model is now available in the model catalog, and the pipeline we are going to create next will be able to use it through its `predict_mutate` procedure.

Training and predicting with a pipeline

Similarly to what has been done in the preceding section, we can configure a pipeline for node classification. Here, I'll just show you the parts that need to be updated, which are actually quite simple since we just need to modify the pipeline's features: we are not going to use the Louvain community or the number of movies anymore. Instead, we want to use the embeddings generated with the GraphSAGE model we just trained. This is achieved with the following feature configuration:

```
pipe.addNodeProperty(
     "beta.graphSage",
     modelName="model-gs-clf-emb",    # the name of the model
we've trained for GraphSAGE
     mutateProperty="embedding"  # the name of the feature
saved in the projected graph
)
pipe.selectFeatures(["embedding"])  # we use the new feature to
train our model
```

The pipeline will take care of computing embeddings for our train and test nodes at the prediction stage.

Also, note that contrary to what we have done in the previous section, we have to use the same node labels in the projected graphs, for training and testing, which is due to how GraphSAGE manages node labels. So, our projected graph for training is defined by the following:

```
projected_graph_object = create_projected_graph(
     gds,
     graph_name="graph-clf-emb",
     node_spec={
          "Person": {
               "label": "MainTrainDownSampled",
               "properties": {
                    "nbMovies": {"defaultValue": 0},
                    "isUSCitizen": {},
               }
          },
     },
     relationship_spec={"KNOWS": {"orientation": "UNDIRECTED",
"aggregation": "SINGLE"}},
)
```

In comparison, our projected graph for testing must be defined by the following:

```
projected_graph_object = create_projected_graph(
    gds,
    graph_name="graph-clf-emb-test",
    node_spec={
        "Person": {
            "label": "MainTest",
            "properties": {
                "nbMovies": {"defaultValue": 0},
                "isUSCitizen": {},
            }
        },
    },
    relationship_spec={"KNOWS": {"orientation": "UNDIRECTED",
"aggregation": "SINGLE"}},
)
```

These are the only differences to take into account when dealing with an inductive algorithm instead of a transductive one: making sure we have a clear separation between the training and the testing sets so that the testing set does not influence and bias the training phase.

Summary

In this chapter, you learned about how to use GDS pipelines to simplify the processes of training an ML model involving graph-based features. GDS pipelines can be configured to run graph algorithms such as the Louvain algorithm and use the result as a feature in a classification or regression model. These models are part of the GDS, so we do not have to explicitly extract data from Neo4j and use another ML library. Everything can be run using the projected graph, which is stored in the model and pipeline catalogs, and used to make predictions on unseen nodes. This lets us use a single tool to compute graph features and perform ML tasks, including the training and prediction of different models, without explicit data exchange from and to the database.

Additionally, we played with the embedding algorithms included in the GDS, starting to surface their advantages and disadvantages.

In the next chapter, we will use another type of pipeline from the GDS to solve another kind of graph problem: link prediction.

Further reading

If you want to learn more about the topics covered in this chapter, I recommend the following readings:

- *GDS Pipeline documentation*: `https://neo4j.com/docs/graph-data-science/current/machine-learning/node-property-prediction/nodeclassification-pipelines/node-classification/`

Exercise

1. Use a Cypher projection to build the projected graph we used in the first section. It must include nodes with the `MainTrain` label and the `nbMovies` and `isUSCitizen` properties, along with relationships of the `KNOWS` type.
2. Create the graph represented in the following figure (same as *Figure 7.4*) in Neo4j. Then, run the Node2Vec algorithm by changing the `p` and `q` parameters and try and understand their behavior:

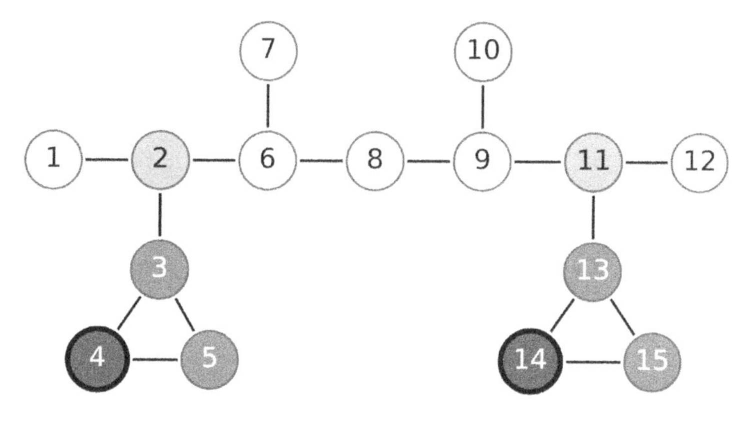

Figure 8.4 – An example graph

9
Predicting Future Edges

Link prediction (**LP**) is a key topic in **Graph Data Science** (**GDS**), since it is a problem very specific to graphs. While we can do classification for many kinds of datasets, not only graphs, LP can only be performed if we have links, meaning if our data is a graph. But the applications of these problems are quite wide: from understanding the dynamics of social network to product recommendations to criminal network analysis.

This chapter is going to give you a short introduction to the LP problem. We will define what observations are and how to build the initial dataset. We will also talk about the metrics that can be used to infer the presence of a hidden or future link and compute them using the GDS library. Finally, we will use a GDS pipeline to build a simple link prediction model, fit it on data stored in Neo4j, and make predictions.

In this chapter, we're going to cover the following main topics:

- Introducing the LP problem
- LP features
- Building an LP pipeline with GDS

Technical requirements

In order to be able to reproduce the examples given in this chapter, you'll need the following tools:

- Neo4j 5.x installed on your computer (see the installation instructions in *Chapter 1, Introducing and Installing Neo4j*)
    - The GDS plugin (*version >= 2.2*)
- A Python environment with Jupyter to run notebooks
- Any code listed in the book is available in the associated GitHub repository,`https://github.com/PacktPublishing/Graph-Data-Science-with-Neo4j`, in the corresponding chapter folder

> **Code samples**
>
> *Unless otherwise indicated, all code snippets in this chapter and the following ones use the GDS Python client.* Library import and client initialization are omitted in this chapter for brevity, but a detailed explanation can be found in *Chapter 6, Building a Machine Learning Model with Graph Features*, in the *Introducing the GDS Python client* section. Also note that the code in the code bundle provided with the book is fully runnable and contains all imports.

Introducing the LP problem

Let's pause for a minute and understand what exactly LP is and how we can formulate this kind of problem using **machine learning** (**ML**) vocabulary.

LP examples

In order to understand what LP is, let's see some real-life scenarios where these problems can be and are used:

- **Social networks**: In a social network containing people who have certain relationships with each other, we can try and predict who the next people to meet or collaborate on a project will be. We can think of the following types of relationships, but there are many more:
  - Social media (know, follow)
  - Communication network (phone call)
  - Research paper authors: co-authorship of a research paper (research collaboration)
- **Criminal networks**: A criminal network, by nature, is not fully known to the people analyzing it (police authorities). The LP technique helps in identifying unknown links between people and better predicting criminal behavior.
- **Entity resolution**: Sometimes, networks will have several nodes linked to the same real entity. That can be website users using different email addresses by mistake or criminals using different aliases on purpose. Given some examples of identical entities exist in the graph, LP algorithms can help in identifying more.
- **Push the boundaries of knowledge**: Other unknowns are directly related to the limits of our own knowledge. For instance, protein interactions within a cell are still not fully understood and a protein-protein interaction network will lack part of the information. LP models help in pushing our knowledge forward, predicting which proteins are more likely to interact and guiding researchers.
- **Recommendations**: Any type of link can be involved in LP. The preceding examples involve relationships between the same types of nodes (people in some cases, proteins in others). But we can imagine predicting relationships between different types of entities, in a so-called

heterogeneous graph, for instance, predicting which product a given customer is likely to purchase next, which movie or TV show a Netflix user will watch and like, or the circuit you're going to run or ride in your next trip.

These are only a few examples! Each time you link two entities together in a **knowledge graph** (**KG**), you can imagine what an LP application would look like in the future. One of the advantages of KG, in this case, is that you do not need to have tons of data: KG provides context to help you get started, a very helpful tool for the cold-start problem. In traditional recommendation engines, based on co-purchases items, for instance, a new product just entered into your catalog will have a hard time competing against older products, with a log of co-purchase data. KG, thanks to the use of product context (category, characteristics such as color, and so on) can push forward these products with short existence.

Let's get more concrete and talk a bit about the possibilities offered by the Netflix dataset created earlier in this book.

LP with the Netflix dataset

The following Netflix graph was built in the preceding chapters. I reproduced some relevant parts of its schema in the following diagram:

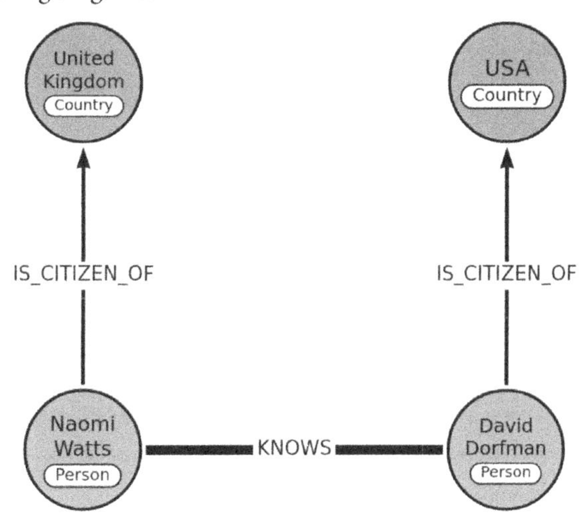

Figure 9.1 – Our Netflix graph schema

The graph contains `Person`, `Country`, and `Movie` nodes. Persons are connected through a `KNOWS` relationship when they have collaborated on at least one movie. The `KNOWS` relationship is persisted in the database, so that we do not have to care about movies anymore. Thanks to public data available on Wikidata, we added the countries of citizenship for each person in the graph, represented by the `IS_CITIZEN_OF` relationship between a person and a country. Let's investigate LP with this dataset in mind.

LP on collaboration graphs

LP is often illustrated with social or co-authorship graphs: nodes of one given label (usually `Person`) are connected with a relationship of one single type (`KNOWS` or `CO_AUTHORED`, for a given research paper, for example), which are examples of **homogeneous** graphs.

Similarly, we could use only the bottom part of the graph represented in *Figure 9.1*, the `Person` nodes and `KNOWS` relationships, to try and predict who will collaborate as actor or director in future movies/TV shows.

However, in the preceding chapter, we did something different: we tried to predict whether people were US citizens or not, using a classification formulation. Can we turn this problem into an LP one?

LP on heterogeneous graphs

Looking at *Figure 9.1*, you can see that the problem can also be framed as an LP problem, where the graph is **heterogeneous** in the sense that it contains different types of nodes (`Person` but also `Country`) and possibly different types of relationships. Even if we are trying to predict a relationship of type `IS_CITIZEN_OF`, the `KNOWS` relationship can be used as contextual information, assuming people who know most US citizens are more likely to be US citizens themselves.

Recommendations are also an example of heterogeneous LP, since they are about predicting links between users and products.

While we know what LP is, we still don't know how to formulate this problem in an ML context. That's the topic of the coming section.

Framing an LP problem

In an LP problem, we are trying to predict whether a link between two given nodes exists or not. That means that *the observations are not nodes anymore*. We could think that the observations are existing links, but they actually form only one part of the dataset: the true (positive) observations, meaning a link exists for observations for which the target is true, the link exists. But to train an ML model, we also need examples of false (negative) observations: links that *do not* exist. That's why I like to think about the observations in an LP model as *pair of nodes*. Take, for instance, the following graph:

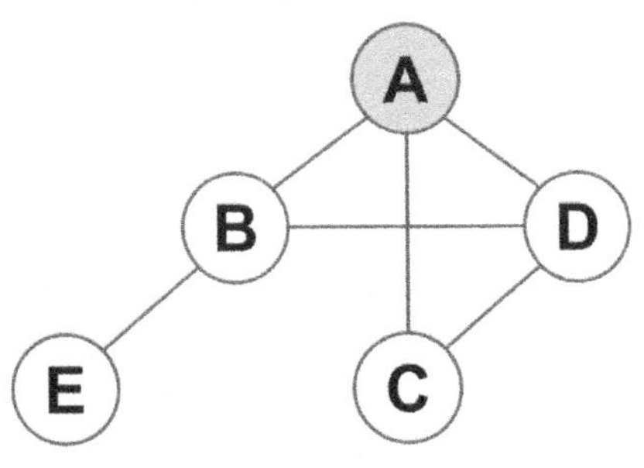

Figure 9.2 – A simple graph

The observations and target required to feed an LP model are as follows:

```
observation_id  target
(A, B)          True
(A, C)          True
(A, D)          True
(A, E)          False
(B, C)          False
(B, E)          True
...
```

The dataset contains both positive samples (existing relationships) and negative ones (pairs of nodes without an existing relationship between them). In this way, an LP problem is a binary classification task: given a pair of nodes, A and E, is there a link likely to appear between these nodes? Or what is the class of the A-E pair: `True` or `False`?

While positive observations are easy to get, negative observations are a bit more challenging: taking into account all non-existing relationships in a graph often leads to very big datasets and a large imbalance between the positive and negative observations. The simplest technique is to randomly sample negative observations, which is the technique used in GDS.

In order to measure the model's quality, you can use the metrics in the binary classification toolbox, for instance, the **receiver operating characteristic** (**ROC**)/**area under the curve** (**AUC**) scores.

Now that we have a better understanding of LP problems in terms of usability and formulation in terms of ML, we are going to focus on the features. In the next section, we are going to get a better understanding of LP problems by investigating what kinds of features can be added to our observations to improve the predictive power of an LP model.

LP features

Here, we'll describe the characteristics that can be attached to a pair of nodes and used as predictors for an LP model. We'll start with topological features, which are built by analyzing both nodes' neighborhoods. Then, we explore how to use each node's features and combine them into a feature vector for the pair.

Topological features

Topological features rely on nodes' neighborhoods and graph topology to infer new links. We can, for instance, use the following:

- **Common neighbors**: Given two nodes, A and B, count the number of common neighbors between A and B. This metric assumes that the more common neighbors A and B have, the more likely they are to be connected.

- **Adamic-Adar**: A variation of the common neighbors approach, the Adamic-Adar metric incorporates the fact that nodes with fewer connections give more information than nodes with many links. In a web page linking hundreds of other pages, the relevance of each link is questionable, compared to a page backlinking only two others.
- **Distance**: The shortest distance between two nodes can also be a good indicator of whether they are likely to be connected. In a social network, my friend's friend is more likely to become a friend than a seventh-degree connection of mine.
- **Same community**: In networks with a clear community structure, we can sometimes assume that two nodes belonging to the same community are more likely to be connected than two nodes in distinct communities.

These metrics, along with some others, are available in GDS, still in the alpha tier at the time of writing. You can, for instance, compute the Adamic-Adar score of two nodes with the following query:

```
MATCH (tom:Person {name: 'Tom Cruise'})
MATCH (leo:Person {name: 'Leonardo DiCaprio'})
RETURN gds.alpha.linkprediction.adamicAdar(tom, leo,
{relationshipQuery: 'KNOWS'}) AS aa_score
```

Note that `adamicAdar` is a function that needs to be called for each pair of nodes in our dataset, whether there is a link between them (positive observations) or not (negative observations).

Defining which topological feature to use depends on your own problem and graph, and the process generating it. A social network might not behave in the same way as a product catalog graph. Deciding which metrics to use requires a deep understanding of your network. You can also function the other way around: use a trial and error method to test several metrics and learn about your network dynamic by finding out which metric has the highest predictive power.

In the next section, we are going to discover another type of feature that can be used in LP: features based on node properties.

Features based on node properties

Each node in a graph may come with its own properties. In the Netflix dataset, we have the title of each movie and the names and number of movies of each person. We have also discovered how to learn features from the graph structure using node embedding algorithms. But these are node features, and remember that we are interested in pairs of nodes in LP. How do we take advantage of node features in this context?

The solution is to combine the node features using a specific function, as illustrated in the following diagram:

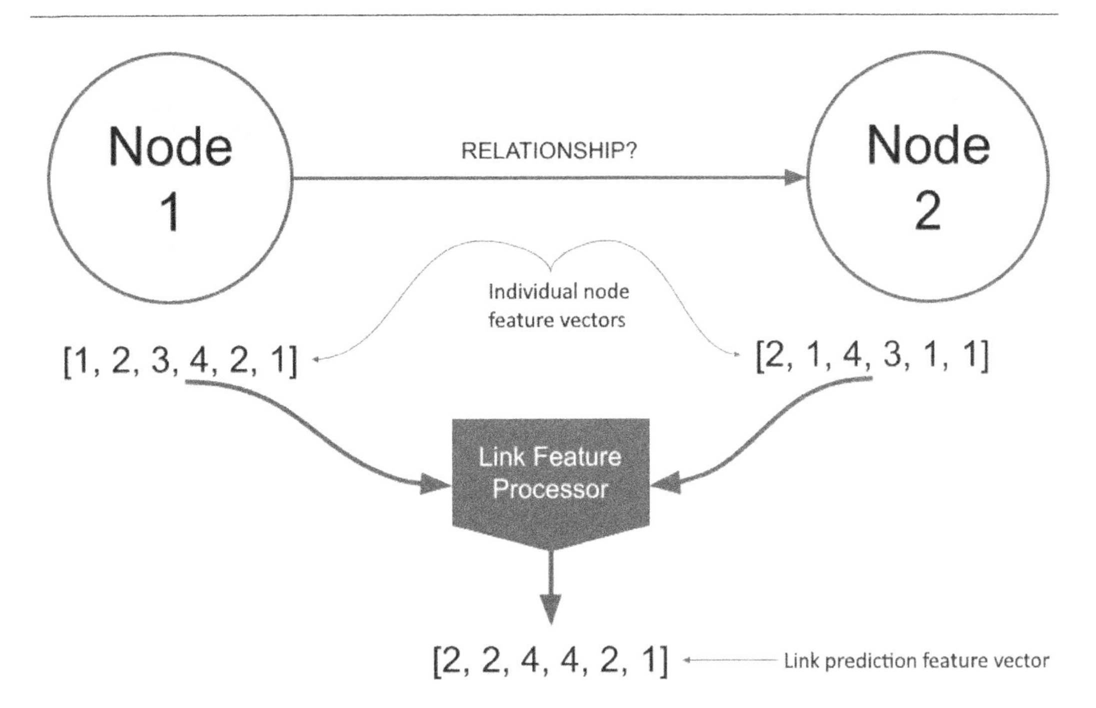

Figure 9.3 – The features of an LP pipeline are based on node features, combined in a link feature processor

Node 1 and **Node 2** have their own feature vectors, which get combined into a single vector by the link feature processor. The LP feature vector is associated with the couple (**Node 1** and **Node 2**) and can be used in an LP model. In this example, the LP vector is built such that its ith element is the maximum of the ith elements of the two input vectors.

There are different types of feature processors. GDS (version 2.3) supports four of them:

- **Same category**: A feature of dimension 1, whose value is 1 if both nodes belong to the same category and 0 otherwise. It can be used as the *same community* feature we introduced in the preceding subsection (*Topological features*)
- **Cosine**: A dimension 1 feature returning the cosine similarity between the two nodes' vectors.
- **L2**: The element-wise squared difference of nodes' features. The resulting vector has the same size as the node's feature vectors.
- **Hadamard**: The element-wise multiplication of the node features vectors, again of the same size as the input vectors.

  Each of these features can be used independently or combined to create a larger input vector.

It is now time for some practice. Using our Netflix dataset, we will build an LP pipeline using the GDS toolbox.

Building an LP pipeline with the GDS

Our task will be to predict the future collaboration of actors and directors, using the homogeneous graph made of `Person` nodes and `KNOWS` relationships. We will only use the persons in the main component according to the connected component algorithm, identified by the `MainComponent` label.

Creating and configuring the pipeline

The process of creating, training, and making predictions with a GDS pipeline is very similar to the node classification case. We will detail the steps in the following subsections.

Building the projected graph

First, we are going to create a projected graph, as follows:

```
projected_graph_object = create_projected_graph(
    gds,
    graph_name="graph-lp-collab",
    node_spec={
        "Person": {
            "label": "MainComponent",
        },
    },
    relationship_spec={
        "KNOWS": {
            "type": "KNOWS_TRAIN",
            "orientation": "UNDIRECTED",
            "properties": {"weight": {"defaultValue": 1.0}}
        }
    },
)
```

It includes nodes with the `MainComponent` label (relabeled `Person` in the projected graph – this is not mandatory) and relationships of type `KNOWS_TRAIN`, also relabeled `KNOWS_TRAIN`. This last relationship type was created with the following Cypher queries, letting apart some relationships for testing:

```
MATCH (p1:MainComponent)-[r:KNOWS]->(p2:MainComponent)
WITH p1, p2, r, CASE WHEN rand() > 0.1 THEN "KNOWS_TRAIN" ELSE
"KNOWS_TEST" END as newType
```

```
CALL apoc.create.relationship(p1, newType, r, p2) YIELD rel
RETURN count(rel)
```

The query leverages APOC to create relationships with a dynamic type (stored in a variable).

> **Relationship direction**
>
> LP pipelines in GDS only support *undirected* relationships.

Now that our projected graph is created and stored in memory, let's configure our pipeline.

Selecting node features

We are going to use node embeddings as input node features for our LP model and use a link feature processor to turn them into a link features:

1. Our first task is to compute these embeddings. We will use the Node2Vec algorithm, and store the computed vector as a new node property in the projected graph:

```
gds.beta.node2vec.mutate(
    projected_graph_object,
    mutateProperty="embedding",
    embeddingDimension=32,
)
```

2. We can then create the `linkPrediction` pipeline:

```
pipe, pipe_info = gds.beta.pipeline.linkPrediction.
create("pipe-lp-collab")
```

3. Add a link feature. Here, we will use only the `hadamard` feature processor:

```
pipe.addFeature("hadamard", nodeProperties=["embedding"])
```

It will take the `embedding` property and compute a new vector that will be used by the model.

4. Then, we add the models to be trained and scored:

```
pipe.addLogisticRegression()
```

If we add multiple models, GDS will train and score each model and only store the best-performing one.

Our pipeline is now ready to be trained.

Pipeline training and testing

Before training the pipeline, let's investigate another nice feature of GDS: its ability to estimate memory requirements for each algorithm and model before it even starts. In this way, we can update our configuration accordingly, without surprises.

Memory usage estimation

When running on large graphs, or moderate-size graphs but using algorithms with high space complexity (for instance, node similarity algorithms, that have $O(n^2)$ space complexity, n being the number of nodes), we may run into more troubles, due to the algorithm requiring more memory than we offer it.

To prevent these issues, we can use the estimate method for almost all algorithms in GDS, including pipelines. With the Python client, estimating the memory requirements of a pipeline is done with the following function:

```
pipe.train_estimate(
    projected_graph_object,
    targetRelationshipType="KNOWS",
    modelName="model-lp-collab",
)
```

It gives us the required memory to train the pipeline. We can use this information to update the Neo4j setting (the `neo4j.conf` file):

```
server.memory.heap.max_size=2G
```

We can even use it to upgrade our hardware configuration so that our process runs smoothly without `OutOfMemoryError`.

Training and testing

The pipeline has been configured, and the memory has been updated if needed, so we can now proceed to train the pipeline. The syntax is similar to the node classification pipeline, with some extra arguments:

```
model, train_result = pipe.train(
    projected_graph_object,
    targetRelationshipType="KNOWS",
    modelName="model-lp-collab",
    randomSeed=42
)
```

The highlighted parameter, `targetRelationshipType`, is not mandatory in our case, because our projected graph contains only one type of relationship. However, I wanted to mention it as it is very important when dealing with heterogeneous graphs, and we are targeting only one type of relationship (think about the `IS_CITIZEN_OF` use case we introduced in the *LP with the Netflix dataset* section).

In the prediction stage, we use the same projected graph (which does not contain the unknown relationships) and ask the algorithm to return the top *K* (`K=5` in the following example) future relationships for each node:

```
res = model.predict_stream(
    projected_graph_object,
    topK=5,
)
```

The returned dataframe (see the following screenshot) contains three columns: **node1**, **node2**, and the predicted probability of the relationship between these two nodes:

	node1	node2	probability
0	9047	21979	0.500482
1	9047	43523	0.500397
2	9047	43525	0.500395
3	9047	43573	0.500382
4	9047	43524	0.500381

Figure 9.4 – LP pipeline predictions

The **node1** and **node2** columns are the GDS internal node IDs, which need to be translated to actual Neo4j nodes, using the `gds.util.asNodes` function.

As you can see by digging deeper into the result dataframe, `res`, the algorithm returns the top five links for each node in the testing set (the `node1` column). There is another approach you can take for prediction, which is to return all links whose probability is higher than a certain threshold instead. For instance, the following code returns only pairs whose estimated probability is higher than `0.7`:

```
res = model.predict_stream(
    projected_graph_object,
    threshold=0,7,
)
```

You can't mix both parameters, but since the `predict_stream` method is able to return the probabilities, it is straightforward to extract the top 10 but only if the probability is higher than a certain threshold thanks to some dataframe manipulation.

Summary

In this chapter, you have learned about the LP problem, an ML technique that's only possible with graph data. It can be used in many contexts to predict future or unknown links between any type of nodes, as long as we have some example or context data. You have learned how to build an LP pipeline with Neo4j's GDS, which takes care of negative observation sampling, model training, and storage for us.

This chapter is the last one where we will talk about predictions and ML. Overall, we have studied several use cases for ML on graphs, including node classification and future/unknown LP. You have learned how to extract graph-based features or embeddings to feed an ML model in your preferred library (we've used scikit-learn). You have also learned that the whole ML pipeline can be managed within Neo4j and its GDS library thanks to built-in pipelines and models.

GDS contains many interesting tools, but it is generally still young compared to other ML tools, such as scikit-learn (which has much more algorithms and options available) and **pyTorch geometric** (**PyG**), a PyTorch extension to deal with graphs, especially graph neural networks. For more advanced applications, requiring the use of different types of algorithms than the ones included in GDS, it might still be required to extract data from Neo4j. But as GDS evolves and covers more use cases, it will become less and less relevant.

We still have one chapter to go before the end of the book. In the next chapter, we will focus on a totally different topic: how to extend GDS to write our own graph algorithm. We will implement our own version of the PageRank algorithm to define node importance, in such a way that it runs on a projected graph and is usable from Cypher, like any other GDS algorithm.

Further reading

If you want to learn more about the topics covered in this chapter, I recommend the following resources:

- An LP analysis with topological metrics and `scikit-learn` is presented in the book *Graph Algorithms* by M. Needham and A. Hodler, O'Reilly (*Chapter 8, Building a GDS Pipeline for Node Classification Model Training*).
- This paper introducing LP problems: *Link Prediction in Complex Networks: A Survey* by L. Lu and T. Zhou: `https://arxiv.org/abs/1010.0725`.
- Some more complex LP examples:
  - LP on heterogeneous graphs:
    - Using PyG: *Link Prediction on Heterogeneous Graphs with PyG* by J. Eric Lenssen and M. Fey: `https://medium.com/@pytorch_geometric/link-prediction-on-heterogeneous-graphs-with-pyg-6d5c29677c70`
    - Using GraphSAGE for recommendations in heterogeneous graphs: *Graph Neural Networks: Link Prediction (Part II)* by L. Faik: `https://medium.com/data-from-the-trenches/graphical-neural-networks-link-prediction-part-ii-c60f6d97fd97`
  - Multi-class link prediction to predict the link type: *Knowledge graph completion with PyKEEN and Neo4j* by T. Bratanic: `https://towardsdatascience.com/knowledge-graph-completion-with-pykeen-and-neo4j-6bca734edf43`. It is also a nice illustration of how to export data from Neo4j to another library, here *PyKEEN*, a Python package for computing graph embeddings.

10

Writing Your Custom Graph Algorithms with the Pregel API in Java

In this final chapter related to creating data science projects on graphs using Neo4j and its plugins, we are going to use an advanced feature of GDS: the Pregel API. This API lets us use the optimized in-memory projected graph to run an algorithm written in Java. GDS takes care of everything else, including parallelism and how to return the result (stream or write back to Neo4j). We will use the PageRank algorithm as an example and learn about its principles before studying a small Python implementation. Then, we will implement it with the Pregel API and test our algorithm with the GDS tools. Finally, we will build the JAR file needed to run our algorithm from Cypher, like any other GDS algorithm.

In this chapter, we're going to cover the following main topics:

- Introducing the Pregel API
- Implementing the PageRank algorithm
- Testing our code
- Using our algorithm from Cypher

Technical requirements

To be able to reproduce the examples given in this chapter, you'll need the following tools:

- Neo4j 5.x installed on your computer (see the installation instructions in *Chapter 1, Introducing and Installing Neo4j*).
    - The GDS plugin (*version >= 2.2*).
- Java (OpenJDK 11). We will also use the gradle build tools.
- It is advised that you use an IDE to manage dependencies and build projects in Java.

Introducing the Pregel API

Neo4j allows you to write plugins: following some API, you can write code in Java, which can then be used from Cypher through, for instance, the `CALL` statement, given that the JAR file containing your code has been placed in the `plugins` folder and your code has been properly annotated so that Neo4j can find the relevant information. That's how APOC and GDS are implemented. Thanks to the Pregel API, we can extend not only Neo4j but GDS itself, leveraging its main features.

In this section, we will cover these GDS features and the basic principles behind the Pregel API.

GDS's features

GDS allows users to extend it while taking advantage of many common functionalities, such as the following:

- **In-memory projected graph**: We won't have to write code to create a projected graph – we can directly work on an existing projected graph in the GDS graph catalog.
- **Stream/write/mutate procedures**: The execution modes are automatically created for the same algorithm.
- **Algorithm configuration**: Here, we can define the required configuration parameters and parse user input against these requirements. This functionality also offers already-created configuration classes containing common parameters such as the number of concurrent threads to control parallel jobs, write property names, and more.

The exposed API from GDS follows a standard called **Pregel**. We are going to introduce it in the next section.

The Pregel API

This method was introduced in the paper *Pregel: a system for large-scale graph processing* in 2010 by G. Malewicz et al. from Google. Its essence is to "*enable large-scale graph data analytics.*"

Though invented by Google, this method has been used and implemented by several other vendors. You can find similar work in ArangoDB (another database with a graph component and analytics) or Apache Spark GraphX (Spark's API for graph computations), for instance.

It relies on a few concepts:

- **Message passing** between neighbors. This means we have to define the following:
  - Which message is *sent* from one node to its neighbors.
  - What a given node should do with the messages it *receives*, usually via some kind of aggregation. This part of the process can, in theory, be processed in batches to speed up the whole process.

  These two steps are performed for each node in the graph in what is called a **superstep**. Supersteps are repeated until a stop criterion is met.

- **Stop criteria**: A step criterion specifies when this message propagation will stop. By default, it runs until a certain number of iterations has been reached, or until there are no more messages to be sent.

Pregel also has the concept of active/inactive nodes:

- In the beginning, all nodes are active.
- At some point, depending on the algorithm, a node can decide it has reached some stability and deactivate itself. In this case, it won't send messages anymore.
- This deactivation works until the node receives new messages from its neighbors, in which case it has to update its value and choose whether to be activated or not.

This concept is called **vote to halt**: each node can vote to deactivate itself. We will not focus too much on this concept here because we are not going to use it in the example algorithm for this chapter, but it is quite powerful. Refer, for instance, to the example given in the Pregel paper referenced at the beginning of this section.

In the rest of this chapter, we will implement an algorithm with this API and use it within Neo4j. To get started, we are going to introduce this algorithm and understand it by covering a simple Python implementation.

> **Note**
>
> For the next three sections, a running Neo4j database is *not* needed. We will implement and test code within our Java environment, in total isolation.

Implementing the PageRank algorithm

As an example, we will use the PageRank algorithm. It is a centrality metric developed by Larry Page, Google's co-founder, to rank results on the search engine.

In this section, we will dig into the algorithm's mechanisms and work on a simple implementation using Python before implementing the algorithm in Java, leveraging the Pregel API.

The PageRank algorithm

This algorithm is based on the following assumptions:

1. The more connections you have, the more important you are.
2. Not all connections share the same weight. For example, let's say a backlink from the New York Times is driving more traffic to your website than a backlink from a less popular website. Scores are propagated from neighbors to account for the neighbor's importance.
3. At the same time, links from a website with fewer links show more relevance. Imagine that there's a Wikipedia article linking every single noun to the corresponding page. In this context, links would carry weak information because they are just linking irrelevant words, compared to how it works now, where only terms relevant to the current item are linked: scores are normalized by the number of outgoing links.

Mathematically speaking, this is translated into the following formula:

$$PR(i) = (1 - d) + d \times \sum_{n} \frac{PR(n)}{L(n)}$$

Here, we have the following:

- *PR(i)* is the PageRank score of node *i*
- *n* denotes the neighbors of node *i* (model points 1 and 2)
- *L(n)* is the outgoing degree of node *n* (model point 3)
- *d* is a **damping factor** that's used to model the fact that, at some point, users will not follow links but arrive at a new website by direct means

To understand the algorithm even further, let's implement it with Python.

Simple Python implementation

PageRank is an iterative algorithm, where at iteration N^k, the PR scores of the previous iteration, *k-1*, are used and aggregated from the node's neighborhood. As with many iterative algorithms, it stops if either of the following two conditions is met:

- A certain stability is reached in the scores computed as iteration *k* compared to the previous one, meaning the scores are not evolving anymore within some **tolerance**.
- A **maximum number of iterations** is reached, in which case the algorithm has not converged but is stopped anyway to prevent infinite calculations. It is usually set around a value of 0.85.

Let's go ahead and implement the algorithm in Python. We will use a `networkx` graph object as input, from which we can iterate over nodes and find the node's neighbors.

The code for this section is available in the `PageRank_Python` notebook.

> **Warning**
> This implementation is not optimized at all. Again, it is being shown for educational purposes only. If you want to dig deeper into the topic, you can check the implementation used in the `networkx` package: `https://github.com/networkx/networkx/blob/83948b6dc61f507adb4cd1a71561eac9ed3b72ec/networkx/algorithms/link_analysis/pagerank_alg.py#L10` (last commit at the time of writing).

Here is a possible implementation with Python:

1. First, import the required packages. We will use `networkx` for the graph object's representation. We will also use the `warnings` package to raise a warning if the algorithm has not converged:

```
import warnings
import networkx as nx
```

2. Define a function with four parameters – the input graph, `G`, the damping factor, `dp`, the maximum number of iterations allowed, and the tolerance we want to reach:

```
def page_rank(
    G,
    dp=0.85,
    max_iterations=20,
    tolerance=0.01
):
```

3. Initialize the initial PR values to `1.0` for all nodes:

```
    # initialization
    pr_values = {
        node: 1.0
        for node in G.nodes
    }
```

4. Initialize the variables we will need in the rest of the code:

```
    # degree and number of nodes
    out_degree = G.out_degree()
    number_of_nodes = G.number_of_nodes()
```

5. Now, start iterating. At each iteration, we will save the previously computed PR value in a new variable called `old_pr_values` and initialize a new `pr_values` variable with the fixed `(1-dp)` part of the formula, which is common to all nodes:

```
iteration_number = 0
while True:
    # copy PR values from previous iteration
    old_pr_values = dict(pr_values)
    # PR values for this iteration,
    # initialized to 1-dp
    pr_values = {
        node: 1 - dp
        for node in G.nodes
    }
```

6. Then comes the message passing part: *each node sends its PR value from the previous step to each of its neighbors:*

```
    # iterate over all nodes
    for node in G.nodes:
        # send message to neighbors
        for neighbor in G.neighbors(node):
            pr_values[neighbor] += (
                dp
                 old_pr_values[node]
                / out_degree[node]
            )
```

7. Finally, we can check the stopping criteria: tolerance and the number of iterations. If the maximum number of iterations is reached without convergence, we raise a warning. In both cases, the `pr_values` list is returned:

```
    # check stop criteria
    convergence = sum(
        abs(pr_values[n] - old_pr_values[n])
        for n in G.nodes
    ) / number_of_nodes
    if convergence < tolerance:
        break
```

```
            iteration_number += 1
            if iteration_number > max_iterations:
                warnings.warn(
                    "Max number of iterations reached"
                )
                break

        return pr_values
```

Now, we can test our function.

8. Create a graph object:

```
G = nx.DiGraph()
G.add_node("A")
G.add_node("B")
G.add_node("C")
G.add_node("D")

G.add_edge("A", "B", weight=1)
G.add_edge("A", "D", weight=3)
G.add_edge("B", "A", weight=2)
G.add_edge("C", "B", weight=1)
G.add_edge("D", "B", weight=4)
```

This graph contains only 4 nodes and 5 edges and is represented in the following diagram:

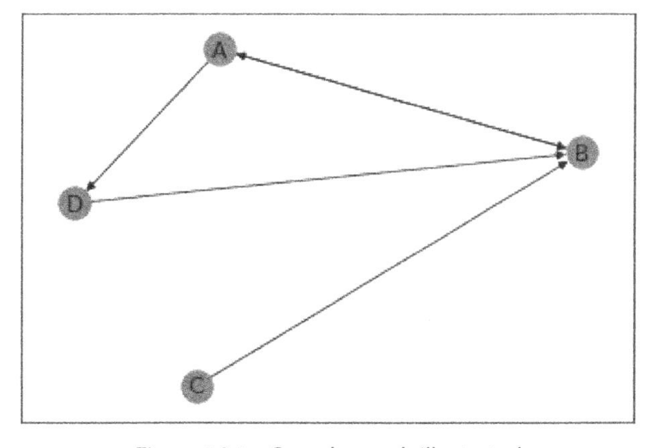

Figure 10.1 – Sample graph illustrated

9. Call our new function:

```
page_rank(G)
```

The (rounded) results for this small graph are as follows:

```
{'A': 1.5,
 'B': 1.6,
 'C': 0.15,
 'D': 0.78}
```

This section has given you an understanding of the algorithm and a better idea of its internal working thanks to a Python implementation. Now, let's learn how to turn this into a GDS plugin using the Pregel API.

Pregel Java implementation

The code for this section is available in the `Chapter10/pregel` folder in this book's code bundle and GitHub repository associated with this book.

To take advantage of the Pregel API exposed by GDS, follow these steps:

1. In Java, everything is a class, so let's start by defining our `PageRank` class in a `PageRank.java` file:

```
public class PageRank implements
PregelComputation<PageRank.PrConfig> {
```

`PageRank.PrConfig` is another class we are going to define inside the main `PageRank` class.

2. Let's implement the `PrConfig` class since we are talking about it:

```
    @ValueClass
    @Configuration("PrConfigImpl")
    @SuppressWarnings("immutables:subtype")
    public interface PrConfig extends
PregelProcedureConfig) {

        // set a default value for max iterations
       // from PregelProcedureConfig
        @Value.Default
        default int maxIterations() {
            return 10;
        }
```

```
        // declare new config parameter with a default
value
        @Value.Default
        default double dampingFactor() {
            // default value
            return 0.85;
        }

        static PrConfig of(CypherMapWrapper userInput) {
            return new PrConfigImpl(userInput);
        }
    }
```

In the preceding code block, we are defining a class that extends the `PregelProcedureConfig` class, which already contains some mandatory configuration parameters for the Pregel API, such as `maxIterations`. However, here, we are doing the following:

- Adding a default value for the `maxIterations` parameter so that it does not have to be set manually each time we use the algorithm
- Defining another parameter, the damping factor, which is specific to the PageRank algorithm, and setting it at a default value of `0.85`, which is the most common value

3. Still within the `PageRank` class, we must define the node schema, which defines the type of information we want to store for each node during the computation, as well as sends the results:

```
@Override
public PregelSchema schema(PrConfig config) {
    return new PregelSchema.Builder()
            .add(PR_KEY, ValueType.DOUBLE)
            .build();
}
```

In our case, we just need one field for each node – the current PR value – which we are going to save as a `Double` number. In the next section (*Implementing the tolerance-stopping criteria*), we will look at another example where we need to store multiple values.

4. The next step is to initialize the algorithm, which we can do by setting a default value of `1.0` for the PR of each node:

```
@Override
public void init(InitContext<PrConfig> context) {
```

```
        context.setNodeValue(PR_KEY, 1.0);
    }
```

Finally, we can implement the compute method, which is where the algorithm is implemented.

5. Define the method with two arguments – the current node context and the messages it receives:

```
    @Override
    public void compute(ComputeContext<PrConfig> context,
Messages messages) {
```

6. Get prValue from the context (previous iteration):

```
        double prValue = context.doubleNodeValue(PR_KEY);
```

7. Compute new ranks based on the neighbor's ranks (messages received):

```
            double oldPrNeighbors = 0;
            for (var message : messages) {
                oldPrNeighbors += message;
            }
            var dp = context.config().dampingFactor();
            prValue = (1 - dp) + dp * oldPrNeighbors;
            context.setNodeValue(PR_KEY, prValue);
```

Once this new prValue has been computed, we must save it into the node's context.

There is a subtlety here compared to the Python implementation: since we are *aggregating received messages* (and not sent messages, as we did in Python), we do not have access to the output degree of the node sending the message, just its message. Due to this, *the output degree is sent together with the PR value in the message* (see the next step).

8. Once the computation has been done, we can send the message to all node neighbors. This message contains our prValue, already divided by the outgoing node degree of the current node:

```
        context.sendToNeighbors(
            prValue / context.degree()
        );
```

And there we go – our first implementation of PageRank using the Pregel API of GDS is complete!

> **Note**
> The PageRank algorithm is already implemented in GDS. The implementation we are going to work on here is for educational purposes only, due to the simple nature of the algorithm and its perfect fit with the Pregel API. Prefer the production-tested official implementation within GDS. Some differences in the result are also expected.

There is still one major difference compared to our Python implementation: while the number of iterations is taken care of by GDS, we have not talked about the tolerance-stopping criteria. Let's improve our code so that it takes this parameter into account before we start testing.

Implementing the tolerance-stopping criteria

For this, let's implement a new class called `PageRankTol`:

```
public class PageRankTol implements
PregelComputation<PageRankTol.PrTolConfig> {
```

To achieve our goal, we will use another method exposed by the Pregel API:

```
public boolean masterCompute(MasterComputeContext<_> context)
```

This method is called at the end of each superstep so that it will let us compute the sum of differences between the previously computed scores and the new ones. But to do so, we also need to save this difference somewhere, since the old scores are not saved. We will use the schema for this and add a new property:

```
@Override
public PregelSchema schema(PrTolConfig config) {
    return new PregelSchema.Builder()
            .add(PR_KEY, ValueType.DOUBLE)
            .add(PR_DIFF, ValueType.DOUBLE, PregelSchema.
Visibility.PRIVATE)
            .build();
}
```

The `PR_DIFF` variable, which is of the `Double` type, contains extra information: we must set its visibility to private so that it can be used from within our `PageRankTol` class but will not be part of the returned result in Neo4j.

Next, we need to fill this new field during the computation, in the `compute` method:

```
double prValue = context.doubleNodeValue(PR_KEY);
double oldPrValue = prValue;
// rest of the code is unchanged
context.setNodeValue(PR_KEY, prValue);
context.setNodeValue(PR_DIFF, Math.abs(prValue - oldPrValue));
```

Finally, we can implement our `masterCompute` method. For this, we need a helper class to iteratively compute the sum of all differences:

```
private static class PrDiffSum {
    public double sumDelta;
    public PrDiffSum() {
        sumDelta = 0;
    }
}

@Override
public boolean masterCompute(MasterComputeContext<PrTolConfig>
context) {
    PrDiffSum diff = new PrDiffSum();
    context.forEachNode(node_id -> {
            diff.sumDelta += context.doubleNodeValue(node_id, PR_
DIFF);
        return true;
    });
    double tol = context.config().tolerance();
    double sumDeltaNormed = diff.sumDelta / context.
nodeCount();
    // stop if the sum of all differences normalized
    // to number of nodes is lower than the tolerance
    return sumDeltaNormed < tol;
}
```

We will see the difference in action when we test our code, which we are going to do next. We are going to write a unit test using the GDS tools without the need for a Neo4j database.

Testing our code

GDS also provides a utility for writing unit tests for our code. We are going to test both implementations, `PageRank` and `PageRankTol`, starting with the former.

Test for the PageRank class

Let's get started and detail the code block by block:

1. First, we must define our test class, called `PageRankTest`:

```
@GdlExtension
class PageRankTest {
```

2. Then, we must create a graph object from a Cypher string, using `NATURAL` orientation for the relationship (the default one):

```
    @GdlGraph(orientation = Orientation.NATURAL)
     private static final String TEST_GRAPH =
             "CREATE" +
                     "   (A:Node)" +
                     ",  (B:Node)" +
                     ",  (C:Node)" +
                     ",  (D:Node)" +
                     ",  (A)-[:REL]->(B)" +
                     ",  (A)-[:REL]->(D)" +
                     ",  (B)-[:REL]->(A)" +
                     ",  (C)-[:REL]->(B)" +
                     ",  (D)-[:REL]→(B)";

    @Inject
    private TestGraph graph;
```

3. After that, we must run a Pregel job using our test graph and our `PageRank` implementation:

```
@Test
   void runPR() {
       var config = PrConfigImpl.builder()
               .dampingFactor(0.85)
               .isAsynchronous(false)
               .build();
```

```
        var pregelJob = Pregel.create(
                graph,
                config,
                new PageRank(),
                Pools.DEFAULT,
                ProgressTracker.NULL_TRACKER
        );
```

4. Finally, we must compare the results with the expected value for each node:

```
        HugeDoubleArray nodeValues = pregelJob.run().
    nodeValues().doubleProperties(PR_KEY);
     // System.out.println(nodeValues);

     // expected values from our Python implementation
     var expected = new HashMap<String, Double>();
     expected.put("A", 1.4818);
     expected.put("B", 1.5885);
     expected.put("C", 0.15);
     expected.put("D", 0.7797);

     assertDoubleValues(
     graph, nodeValues::get, expected, 1E-3
     );

     }
    }
```

We can do this using the run configuration illustrated in the following screenshot:

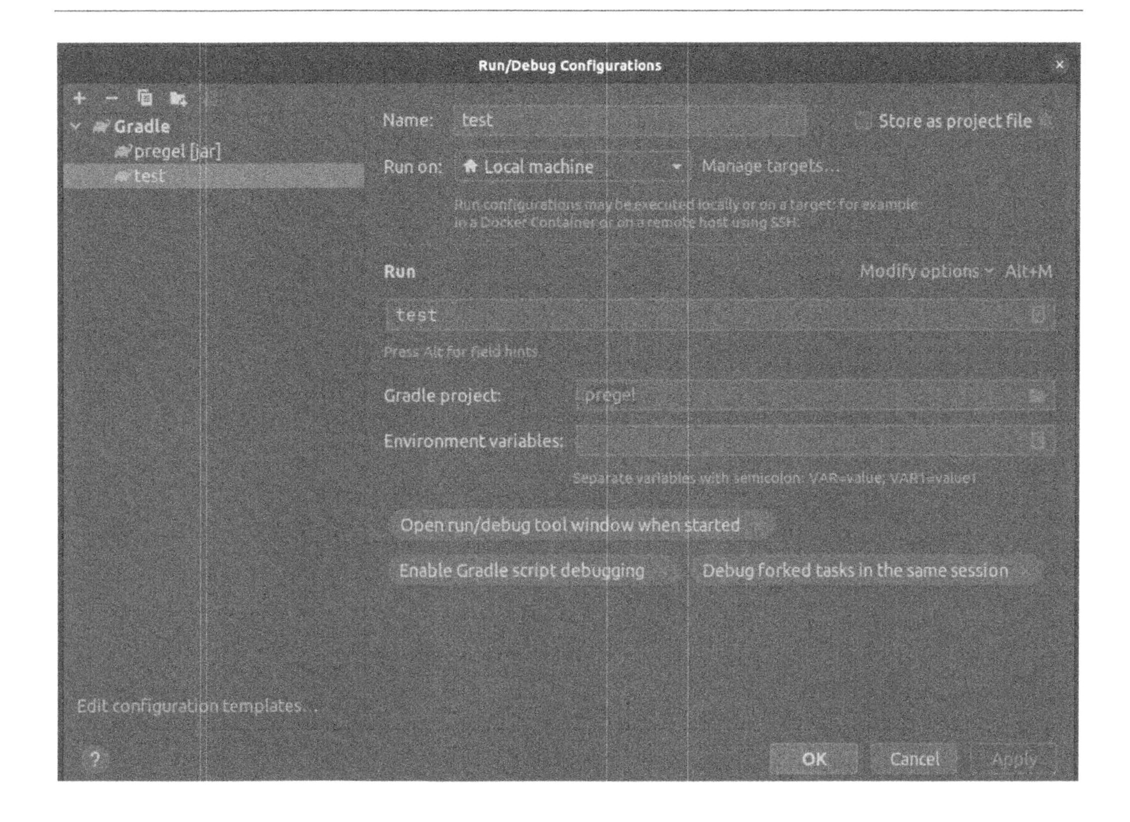

Figure 10.2 – Run configuration for tests in IntelliJ

We will obtain the following result, meaning our test *passed* as expected:

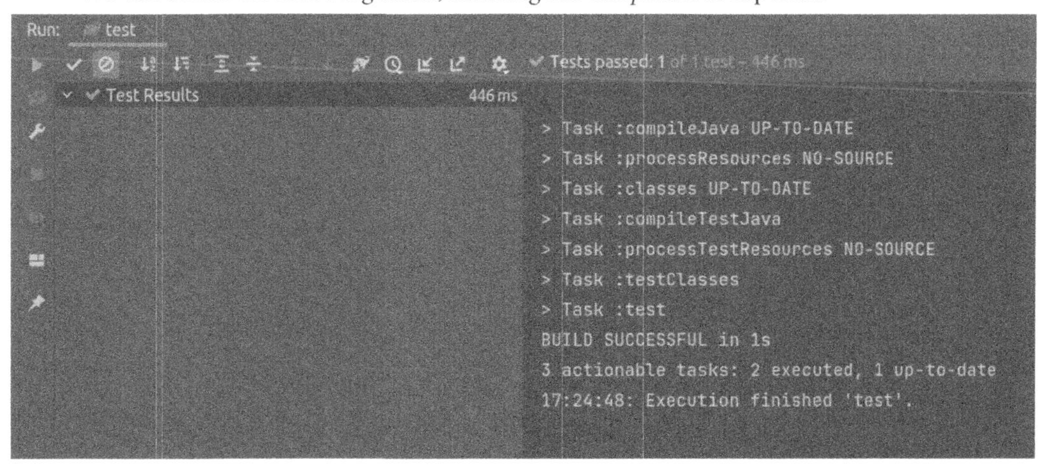

Figure 10.3 – Test result in IntelliJ

Now, let's learn how to update this test for the `PageRankTol` implementation.

Test for the PageRankTol class

We will use the same structure that we used previously to test this class but replace the configuration and algorithm classes with the new ones:

```
class PageRankTolTest {
    // ...
    void runPRTol() {
        var config = PrTolConfigImpl.builder().build();
        var pregelJob = Pregel.create(
                graph,
                config,
                new PageRankTol(),
                Pools.DEFAULT,
                ProgressTracker.NULL_TRACKER
        );

        PregelResult res = pregelJob.run();

        int nbIt = res.ranIterations();
        assert nbIt < config.maxIterations();

        HugeDoubleArray nodeValues = res.nodeValues().
doubleProperties(PR_KEY);
        // ...

    }
}
```

Compared to our previous test class, we are also checking that the algorithm stops before reaching the max number of iterations, which is set to 100 by default. You can rerun the tests on the project and see that they are working fine.

Now that we have programmatically tested our code, we can distribute it so that we can finally use our algorithm from Cypher.

Using our algorithm from Cypher

This is the last step of plugin development: we are going to annotate our file so that Cypher knows what we are talking about, then build the JAR file and test it from Cypher.

Adding annotations

Before generating the JAR file, we need to annotate our `PageRank` class so that we can configure how it can be used from Neo4j:

```
@PregelProcedure(name = "gdsbook.pr", modes = {GDSMode.STREAM,
GDSMode.MUTATE, GDSMode.WRITE})
public class PageRank implements PregelComputation<PageRank.
Config> {
// rest of the code is unchanged ...
}
```

Here, we specify two important parameters:

- **The procedure's name**: This is the name of the procedure known by Cypher. In short, it means that we will be able to write the following:

  ```
  CALL gdsbook.pr.<mode>(...) in Cypher
  ```

- **The procedure's modes**: Here, we can choose one or many among the GDS modes (stream, mutate, write, and stats) that will be available.
- Optionally, we can add a description with `description=""`.

Similarly, update the `PageRankTol` class with a similar annotation:

```
@PregelProcedure(name = "gdsbook.prtol", modes = {GDSMode.
STREAM, GDSMode.MUTATE, GDSMode.WRITE})
public class PageRankTol implements
PregelComputation<PageRankTol.PrTolConfig> {
```

Now, we can build our JAR.

Building the JAR file

To do so, create a new **run configuration** in IntelliJ, as illustrated in the following screenshot:

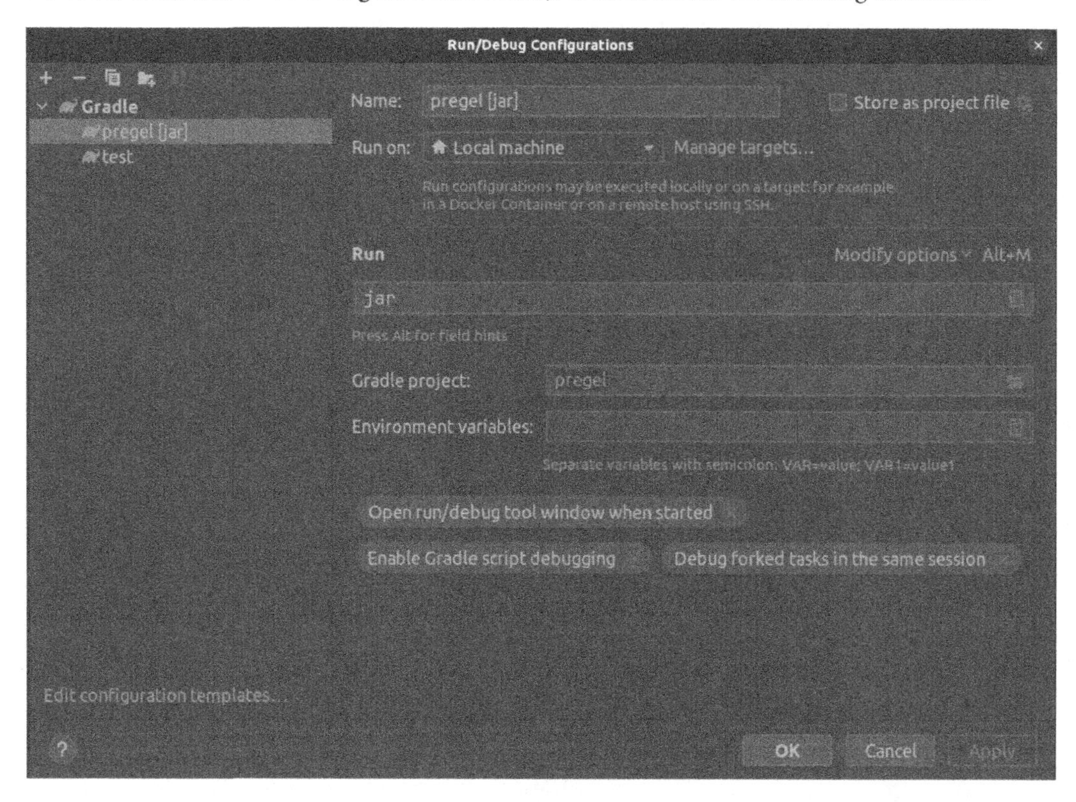

Figure 10.4 – Configuration to build the JAR file in IntelliJ

Here, we are using gradle and using its `jar` run mode. After running this configuration, a new JAR file called `pregel.jar` will be created in `build/libs/` (`pregel` is the name of my local project).

You can copy this file into the plugins folder of a running Neo4j database. In this chapter, I will use the sample graph database we created in *Chapter 7, Automatically Extracting Features with Graph Embeddings for Machine Learning*, which contains around 800 nodes, though you can use any database that contains some data.

Updating the Neo4j configuration

To use user-defined procedures, we need to update the database configuration. To do so, open the configuration file from Neo4j Desktop, as illustrated in the following screenshot:

Figure 10.5 – Opening the settings from Neo4j Desktop

Update the `dbms.security.procedures.unrestricted` field to the following:

```
dbms.security.procedures.unrestricted=jwt.
security.*,gds.*,gdsbook.*
```

That's it! We are now ready to test our procedure on a real Neo4j database.

Testing our procedure

A procedure written with the Pregel API is a GDS procedure and, like any GDS algorithm, it runs on a projected graph. So, the first step is to create a projected graph (the GDS library needs to be installed for your active database). Here, we are going to use the sample graph we created in a previous chapter. The data file and the Cypher query to rebuild it are provided in this chapter's `README`. Once you have a running database, we can create the projected graph:

```
CALL gds.graph.project("pgraph", "Node", "REL")
```

Then, we can use our procedure:

```
CALL gdsbook.pr.stream("pgraph")
```

The result for the first four nodes is illustrated in the following screenshot:

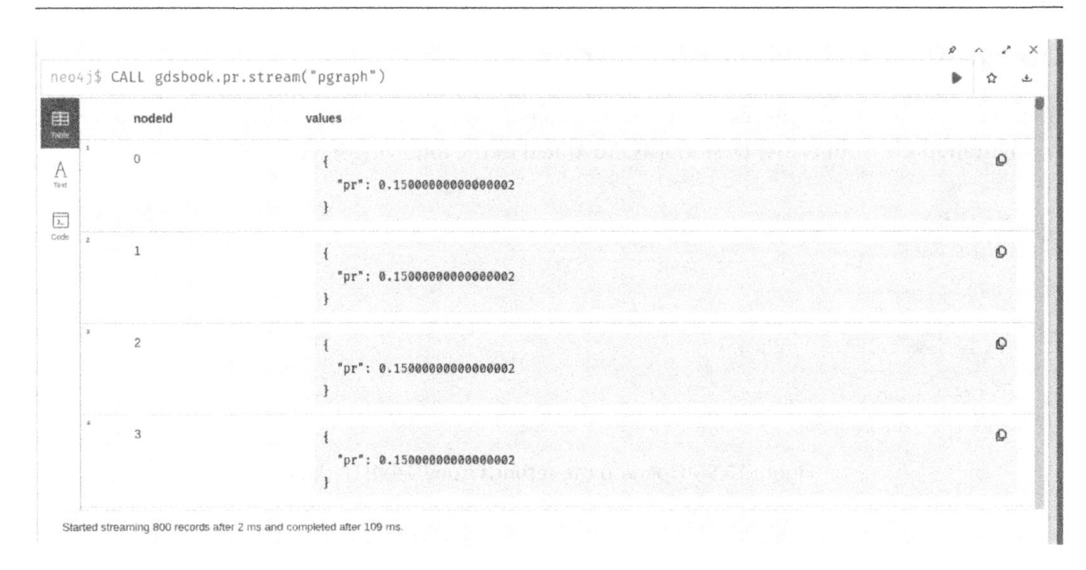

Figure 10.6 – Custom procedure results in Neo4j Browser

As you can see, there are two columns: one containing the GDS internal `nodeId` (from the projected graph) and the other containing the result of our algorithm – a map that, for each node, matches the definition of our `schema` in the `PageRank` class. Our implementation only exposes one field called `pr`, but we could add more. The *Exercises* section at the end of this chapter contains an example of schema with more than one field.

You can also check the result of the `gdsbook.prtol.stream` procedure and observe that the `PR_DIFF` variable we added to our schema with a private scope is not streamed within the result set, as expected.

Summary

This is the end of this chapter, where you were introduced to the method you can use to extend GDS and take advantage of all the common features we are looking for when dealing with graph analytics: memory and CPU performance. The projected graph and GDS internal management of job batches are easily accessible to us if we write a couple of Java classes.

We also studied the PageRank algorithm and implemented two versions of it: one relying only on the maximum number of iterations as stopping criteria, and another version that considers the stability of computed scores compared to the previous iteration, within a certain tolerance. We also learned how to unit test our algorithm by writing a simple test that runs our algorithm on a sample graph, which we were able to define by writing a Cypher `CREATE` statement.

This chapter is also the end of this book! We have come a long way since *Chapter 1, Introducing and Installing Neo4j*, where we introduced the concept of graphs, and *Chapter 2, Using Existing Data to Build a Knowledge Graph*, in which we built our first knowledge graph, which we stored in Neo4j using its Cypher query language. As a data scientist, you can now work with a graph dataset like any other dataset: extracting key metrics to understand it, from its degree distribution to its community structure, if any. You can also train a machine learning model on a graph dataset to classify nodes into several categories or predict the existence of future edges, using both your preferred Python tool and the necessary GDS features, which are the model and pipeline catalog.

In short, this introductory book has given you all the tools you need to build a graph and extract powerful insights from it.

Fortunately, the topics of graph data science and graph machine learning are not limited to the ones covered in this book: graph neural networks, encoder/decoder models, and natural language processing are a few examples of topics you can explore next. Some resources have been provided in the *Further reading* section. Keep learning!

Further reading

If you want to learn more about the topics covered in this chapter, I recommend the following resources:

- I wrote a medium story explaining how to write Neo4j plugins (not with GDS, as we did in this chapter). You can find it here: `https://medium.com/@st3llasia/writing-custom-procedures-for-neo4j-d20e28dc7afe`.
- The Pregel paper: `https://dl.acm.org/doi/abs/10.1145/1807167.1807184`.
- The Pregel API documentation in GDS: `https://neo4j.com/docs/graph-data-science/current/algorithms/pregel-api/`.
- The PageRank algorithm paper: `http://ilpubs.stanford.edu:8090/422/1/1999-66.pdf`. Here are some resources you may find useful to continue your journey with GDS, especially the topics not covered in this book since choices had to be made regarding what to cover:
  - *Knowledge Graphs Applied*, by A. Negro, V. Kus, G. Futia, and F. Montagna (Manning): A lot of topics are covered in this book. While the chapter titles sound like research papers, each is related to a **real use case**. This is a must-read if you want to gain some inspiration about the type of analyses that can be performed on graphs. I especially recommend part 2, which explains how to build a knowledge graph from texts using natural language processing techniques.
  - *Graph Algorithms for Data Science*, by T. Bratanic (Manning): There's a lot of overlap with this book but it's illustrated with concrete examples. I especially recommend the chapters about monopartite/bipartite graphs.

Exercises

Practice with the Pregel API and write an algorithm. If you need some intermediate steps, here are a couple of exercises to help you get started:

1. Update the Python implementation so that it computes the normalized PageRank given by the following formula:

$$PR(i) = (1 - d)/N + d \times \sum_{n} \frac{PR(n)}{L(n)}$$

   Here, N is the total number of nodes in the graph.

   Warning: Be careful with the score initialization.

2. Again, using the Python implementation, take into account relationship weights. *Hint*: Relationship weights are entered into the outgoing degree part of the formula.
3. Update the Java implementation to track the PR values at each step. We want to be able to see the evolution of PR at each iteration when calling the algorithm in Cypher.

   This means adding a new field to our schema of the `double[]` type and extending it at each iteration.

Index

A

B

C

D

E

F

G

H

I

J

Q

R

S

T

U

V

W

Z

Other Books You May Enjoy

If you enjoyed this book, you may be interested in these other books by Packt:

Modern Time Series Forecasting with Python

Manu Joseph

ISBN: 978-1-80324-680-2

- Find out how to manipulate and visualize time series data like a pro
- Set strong baselines with popular models such as ARIMA
- Discover how time series forecasting can be cast as regression
- Engineer features for machine learning models for forecasting
- Explore the exciting world of ensembling and stacking models
- Get to grips with the global forecasting paradigm
- Understand and apply state-of-the-art DL models such as N-BEATS and Autoformer
- Explore multi-step forecasting and cross-validation strategies

Production-Ready Applied Deep Learning

Tomasz Palczewski, Jaejun (Brandon) Lee, Lenin Mookiah

ISBN: 978-1-80324-366-5

- Understand how to develop a deep learning model using PyTorch and TensorFlow
- Convert a proof-of-concept model into a production-ready application
- Discover how to set up a deep learning pipeline in an efficient way using AWS
- Explore different ways to compress a model for various deployment requirements
- Develop Android and iOS applications that run deep learning on mobile devices
- Monitor a system with a deep learning model in production
- Choose the right system architecture for developing and deploying a model

Packt is searching for authors like you

If you're interested in becoming an author for Packt, please visit authors.packtpub.com and apply today. We have worked with thousands of developers and tech professionals, just like you, to help them share their insight with the global tech community. You can make a general application, apply for a specific hot topic that we are recruiting an author for, or submit your own idea.

Share Your Thoughts

Now you've finished *Graph Data Science with Neo4j*, we'd love to hear your thoughts! Scan the QR code below to go straight to the Amazon review page for this book and share your feedback or leave a review on the site that you purchased it from.

https://packt.link/r/1-804-61274-X

Your review is important to us and the tech community and will help us make sure we're delivering excellent quality content.

Download a free PDF copy of this book

Thanks for purchasing this book!

Do you like to read on the go but are unable to carry your print books everywhere? Is your eBook purchase not compatible with the device of your choice?

Don't worry, now with every Packt book you get a DRM-free PDF version of that book at no cost.

Read anywhere, any place, on any device. Search, copy, and paste code from your favorite technical books directly into your application.

The perks don't stop there, you can get exclusive access to discounts, newsletters, and great free content in your inbox daily

Follow these simple steps to get the benefits:

1. Scan the QR code or visit the link below

`https://packt.link/free-ebook/9781804612743`

2. Submit your proof of purchase
3. That's it! We'll send your free PDF and other benefits to your email directly

Made in the USA
Middletown, DE
23 November 2023